CAMBRIDGE LIBRARY COLLECTION

Books of enduring scholarly value

Philosophy

This series contains both philosophical texts and critical essays about philosophy, concentrating especially on works originally published in the eighteenth and nineteenth centuries. It covers a broad range of topics including ethics, logic, metaphysics, aesthetics, utilitarianism, positivism, scientific method and political thought. It also includes biographies and accounts of the history of philosophy, as well as collections of papers by leading figures. In addition to this series, primary texts by ancient philosophers, and works with particular relevance to philosophy of science, politics or theology, may be found elsewhere in the Cambridge Library Collection.

Collected Essays

Known as 'Darwin's Bulldog', the biologist Thomas Henry Huxley (1825–95) was a tireless supporter of the evolutionary theories of his friend Charles Darwin. Huxley also made his own significant scientific contributions, and he was influential in the development of science education despite having had only two years of formal schooling. He established his scientific reputation through experiments on aquatic life carried out during a voyage to Australia while working as an assistant surgeon in the Royal Navy; ultimately he became President of the Royal Society (1883–5). Throughout his life Huxley struggled with issues of faith, and he coined the term 'agnostic' to describe his beliefs. This nine-volume collection of Huxley's essays, which he edited and published in 1893–4, demonstrates the wide range of his intellectual interests. Volume 2 examines the criticism and controversy surrounding Darwin's work, and the implications of his ideas on biological science.

Cambridge University Press has long been a pioneer in the reissuing of out-of-print titles from its own backlist, producing digital reprints of books that are still sought after by scholars and students but could not be reprinted economically using traditional technology. The Cambridge Library Collection extends this activity to a wider range of books which are still of importance to researchers and professionals, either for the source material they contain, or as landmarks in the history of their academic discipline.

Drawing from the world-renowned collections in the Cambridge University Library, and guided by the advice of experts in each subject area, Cambridge University Press is using state-of-the-art scanning machines in its own Printing House to capture the content of each book selected for inclusion. The files are processed to give a consistently clear, crisp image, and the books finished to the high quality standard for which the Press is recognised around the world. The latest print-on-demand technology ensures that the books will remain available indefinitely, and that orders for single or multiple copies can quickly be supplied.

The Cambridge Library Collection will bring back to life books of enduring scholarly value (including out-of-copyright works originally issued by other publishers) across a wide range of disciplines in the humanities and social sciences and in science and technology.

Collected Essays

VOLUME 2: DARWINIANA

THOMAS HENRY HUXLEY

CAMBRIDGE UNIVERSITY PRESS

Cambridge, New York, Melbourne, Madrid, Cape Town,
Singapore, São Paolo, Delhi, Tokyo, Mexico City

Published in the United States of America by Cambridge University Press, New York

www.cambridge.org
Information on this title: www.cambridge.org/9781108040525

© in this compilation Cambridge University Press 2011

This edition first published 1893
This digitally printed version 2011

ISBN 978-1-108-04052-5 Paperback

COLLECTED ESSAYS

By T. H. HUXLEY

VOLUME II

DARWINIANA

ESSAYS

BY

THOMAS H. HUXLEY

London

MACMILLAN AND CO.

1893

PREFACE

I HAVE entitled this volume "Darwiniana"
because the pieces republished in it either treat of
the ancient doctrine of Evolution, rehabilitated and
placed upon a sound scientific foundation, since
and in consequence of, the publication of the
"Origin of Species;" or they attempt to meet the
more weighty of the unsparing criticisms with
which that great work was visited for several years
after its appearance; or they record the impression
left by the personality of Mr. Darwin on one who
had the privilege and the happiness of enjoying his
friendship for some thirty years; or they endeavour
to sum up his work and indicate its enduring
influence on the course of scientific thought.

Those who take the trouble to read the first
two essays, published in 1859 and 1860, will, I
think, do me the justice to admit that my zeal to
secure fair play for Mr. Darwin, did not drive me
into the position of a mere advocate; and that,
while doing justice to the greatness of the argu-

ment I did not fail to indicate its weak points. I have never seen any reason for departing from the position which I took up in these two essays; and the assertion which I sometimes meet with nowadays, that I have "recanted" or changed my opinions about Mr. Darwin's views, is quite unintelligible to me.

As I have said in the seventh essay, the fact of evolution is to my mind sufficiently evidenced by palæontology; and I remain of the opinion expressed in the second, that until selective breeding is definitely proved to give rise to varieties infertile with one another, the logical foundation of the theory of natural selection is incomplete. We still remain very much in the dark about the causes of variation; the apparent inheritance of acquired characters in some cases; and the struggle for existence within the organism, which probably lies at the bottom of both of these phenomena.

Some apology is due to the reader for the reproduction of the "Lectures to Working Men" in their original state. They were taken down in shorthand by Mr. J. Aldous Mays, who requested me to allow him to print them. I was very much pressed with work at the time; and, as I could not revise the reports, which I imagined, moreover, would be of little or no interest to any but my auditors, I stipulated that a notice should be prefixed to that effect. This was done; but it did not prevent a considerable diffusion of the little book in this country and in the United States, nor its translation into more than one foreign language. Moreover Mr. Darwin often urged me to revise and expand the lectures into a systematic popular exposition of the topics of which they treat. I have more than once set about the task : but the proverb about spoiling a horn and not making a spoon, is particularly applicable to attempts to remodel a piece of work which may have served its immediate purpose well enough.

So I have reprinted the lectures as they stand, with all their imperfections on their heads. It would seem that many people must have found them useful thirty years ago ; and, though the sixties appear now to be reckoned by many of the rising generation as a part of the dark ages, I am not without some grounds for suspecting that there yet remains a fair sprinkling even of "philosophic thinkers" to whom it may be a profitable, perhaps even a novel, task to descend from the heights of speculation and go over the A B C of the great biological problem as it was set before a body of shrewd artisans at that remote epoch.

T. H. H.

HODESLEA, EASTBOURNE,
April 7th, 1893.

CONTENTS

VII

VIII

IX

X

XI

COLLECTED ESSAYS

VOLUME II

I

THE DARWINIAN HYPOTHESIS

[1859]

THE hypothesis of which the present work of Mr. Darwin is but the preliminary outline, may be stated in his own language as follows :— " Species originated by means of natural selection, or through the preservation of the favoured races in the struggle for life." To render this thesis intelligible, it is necessary to interpret its terms. In the first place, what is a species ? The question is a simple one, but the right answer to it is hard to find, even if we appeal to those who should know most about it. It is all those animals or plants which have descended from a single pair of parents; it is the smallest distinctly definable group of living organisms; it is an eternal and immutable entity; it is a mere abstraction of the human intellect having no existence in nature. Such are a few of the significations attached to

this simple word which may be culled from
authoritative sources; and if, leaving terms and
theoretical subtleties aside, we turn to facts and
endeavour to gather a meaning for ourselves, by
studying the things to which, in practice, the
name of species is applied, it profits us little.   For
practice varies as much as theory.   Let two
botanists or two zoologists examine and describe
the productions of a country, and one will pretty
certainly disagree with the other as to the number,
limits, and definitions of the species into which he
groups the very same things.   In these islands, we
are in the habit of regarding mankind as of one
species, but a fortnight's steam will land us in a
country where divines and savants, for once in
agreement, vie with one another in loudness of
assertion, if not in cogency of proof, that men are
of different species; and, more particularly, that
the species negro is so distinct from our own that
the Ten Commandments have actually no reference
to him.   Even in the calm region of entomology,
where, if anywhere in this sinful world, passion
and prejudice should fail to stir the mind, one
learned coleopterist will fill ten attractive volumes
with descriptions of species of beetles, nine-tenths
of which are immediately declared by his brother
beetle-mongers to be no species at all.

The truth is that the number of distinguishable
living creatures almost surpasses imagination.   At
least 100 000 such kinds of insects alone have been

described and may be identified in collections, and the number of separable kinds of living things is under-estimated at half a million. Seeing that most of these obvious kinds have their accidental varieties, and that they often shade into others by imperceptible degrees, it may well be imagined that the task of distinguishing between what is permanent and what fleeting, what is a species and what a mere variety, is sufficiently formidable.

But is it not possible to apply a test whereby a true species may be known from a mere variety? Is there no criterion of species? Great authorities affirm that there is—that the unions of members of the same species are always fertile, while those of distinct species are either sterile, or their offspring, called hybrids, are so. It is affirmed not only that this is an experimental fact, but that it is a provision for the preservation of the purity of species. Such a criterion as this would be invaluable; but, unfortunately, not only is it not obvious how to apply it in the great majority of cases in which its aid is needed, but its general validity is stoutly denied. The Hon. and Rev. Mr. Herbert, a most trustworthy authority, not only asserts as the result of his own observations and experiments that many hybrids are quite as fertile as the parent species, but he goes so far as to assert that the particular plant *Crinum capense* is much more fertile when crossed by a

distinct species than when fertilised by its proper
pollen ! On the other hand, the famous Gaertner,
though he took the greatest pains to cross the
Primrose and the Cowslip, succeeded only once or
twice in several years; and yet it is a well-
established fact that the Primrose and the Cowslip are only varieties of the same kind of plant.
Again, such cases as the following are well established. The female of species A, if crossed with
the male of species B, is fertile; but, if the female
of B is crossed with the male of A, she remains
barren. Facts of this kind destroy the value of
the supposed criterion.

If, weary of the endless difficulties involved in
the determination of species, the investigator,
contenting himself with the rough practical
distinction of separable kinds, endeavours to
study them as they occur in nature—to ascertain
their relations to the conditions which surround
them, their mutual harmonies and discordancies of
structure, the bond of union of their present and
their past history, he finds himself, according to
the received notions, in a mighty maze, and with,
at most, the dimmest adumbration of a plan.
If he starts with any one clear conviction, it is
that every part of a living creature is cunningly
adapted to some special use in its life. Has not
his Paley told him that that seemingly useless
organ, the spleen, is beautifully adjusted as so
much packing between the other organs? And

yet, at the outset of his studies, he finds that no
adaptive reason whatsoever can be given for one-
half of the peculiarities of vegetable structure.
He also discovers rudimentary teeth, which are
never used, in the gums of the young calf and in
those of the fœtal whale; insects which never
bite have rudimental jaws, and others which
never fly have rudimental wings; naturally blind
creatures have rudimental eyes ; and the halt
have rudimentary limbs. So, again, no animal or
plant puts on its perfect form at once, but all have
to start from the same point, however various the
course which each has to pursue. Not only men
and horses, and cats and dogs, lobsters and
beetles, periwinkles and mussels, but even the
very sponges and animalcules commence their
existence under forms which are essentially
undistinguishable; and this is true of all the
infinite variety of plants. Nay, more, all living
beings march, side by side, along the high road of
development, and separate the later the more like
they are ; like people leaving church, who all go
down the aisle, but having reached the door, some
turn into the parsonage, others go down the
village, and others part only in the next parish.
A man in his development runs for a little while
parallel with, though never passing through, the
form of the meanest worm, then travels for a
space beside the fish, then journeys along with
the bird and the reptile for his fellow travellers :

distribution of animals and plants it seems utterly hopeless to attempt to understand the strange and apparently capricious relations which they exhibit. One would be inclined to suppose *à priori* that every country must be naturally peopled by those animals that are fittest to live and thrive in it. And yet how, on this hypothesis, are we to account for the absence of cattle in the Pampas of South America, when those parts of the New World were discovered ? It is not that they were unfit for cattle, for millions of cattle now run wild there; and the like holds good of Australia and New Zealand. It is a curious circumstance, in fact, that the animals and plants of the Northern Hemisphere are not only as well adapted to live in the Southern Hemisphere as its own autochthones, but are, in many cases, absolutely better adapted, and so overrun and extirpate the aborigines. Clearly, therefore, the species which naturally inhabit a country are not necessarily the best adapted to its climate and other conditions. The inhabitants of islands are often distinct from any other known species of animal or plants (witness our recent examples from the work of Sir Emerson Tennent, on Ceylon), and yet they have almost always a sort of general family resemblance to the animals and plants of the nearest mainland. On the other hand, there is hardly a species of fish, shell, or crab common to the opposite sides of the narrow

distinct from those that now live. Nor is this unlikeness without its rule and order. As a broad
fact, the further we go back in time the less the
buried species are like existing forms ; and, the further apart the sets of extinct creatures are, the less
they are like one another. In other words, there
has been a regular succession of living beings, each
younger set, being in a very broad and general
sense, somewhat more like those which now live.

It was once supposed that this succession had
been the result of vast successive catastrophes,
destructions, and re-creations *en masse;* but
catastrophes are now almost eliminated from
geological, or at least palæontological speculation ;
and it is admitted, on all hands, that the seeming
breaks in the chain of being are not absolute, but
only relative to our imperfect knowledge ; that
species have replaced species, not in assemblages,
but one by one ; and that, if it were possible to
have all the phenomena of the past presented to
us, the convenient epochs and formations of the
geologist, though having a certain distinctness,
would fade into one another with limits as
undefinable as those of the distinct and yet
separable colours of the solar spectrum.

Such is a brief summary of the main truths
which have been established concerning species.
Are these truths ultimate and irresolvable facts,
or are their complexities and perplexities the
mere expressions of a higher law ?

A large number of persons practically assume the former position to be correct. They believe that the writer of the Pentateuch was empowered and commissioned to teach us scientific as well as other truth, that the account we find there of the creation of living things is simply and literally correct, and that anything which seems to contradict it is, by the nature of the case, false. All the phenomena which have been detailed are, on this view, the immediate product of a creative fiat and, consequently, are out of the domain of science altogether.

Whether this view prove ultimately to be true or false, it is, at any rate, not at present supported by what is commonly regarded as logical proof, even if it be capable of discussion by reason ; and hence we consider ourselves at liberty to pass it by, and to turn to those views which profess to rest on a scientific basis only, and therefore admit of being argued to their consequences. And we do this with the less hesitation as it so happens that those persons who are practically conversant with the facts of the case (plainly a considerable advantage) have always thought fit to range themselves under the latter category.

The majority of these competent persons have up to the present time maintained two positions— the first, that every species is, within certain defined limits, fixed and incapable of modification ; the second, that every species was originally produced by a distinct creative act. The second position is obviously incapable of proof or disproof, the direct operations of the Creator not being subjects of science; and it must therefore be regarded as a corollary from the first, the truth or falsehood of which is a matter of evidence. Most persons imagine that the arguments in favour of it are overwhelming; but to some few minds, and these, it must be confessed, intellects of no small power and grasp of knowledge, they have not brought conviction. Among these minds, that of the famous naturalist Lamarck, who possessed a greater acquaintance with the lower forms of life than any man of his day, Cuvier not excepted, and was a good botanist to boot, occupies a prominent place.

Two facts appear to have strongly affected the course of thought of this remarkable man—the one, that finer or stronger links of affinity connect all living beings with one another, and that thus the highest creature grades by multitudinous steps into the lowest; the other, that an organ may be developed in particular directions by exerting itself in particular ways, and that modifications once induced may be transmitted and become hereditary. Putting these facts together, Lamarck endeavoured to account for the first by the operation of the second. Place an animal in new circumstances, says he, and its needs will be altered; the new needs will create new desires, and the attempt to gratify such desires will result in an appropriate modification of the organs exerted. Make a man a blacksmith, and his brachial muscles will develop in accordance with the demands made upon them, and in like manner, says Lamarck, "the efforts of some short-necked bird to catch fish without wetting himself have, with time and perseverance, given rise to all our herons and long-necked waders."

The Lamarckian hypothesis has long since been justly condemned, and it is the established practice for every tyro to raise his heel against the carcase of the dead lion. But it is rarely either wise or instructive to treat even the errors of a really great man with mere ridicule, and in the present case the logical form of the doctrine stands on a very different footing from its substance.

If species have really arisen by the operation of natural conditions, we ought to be able to find those conditions now at work ; we ought to be able to discover in nature some power adequate to modify any given kind of animal or plant in such a manner as to give rise to another kind, which would be admitted by naturalists as a distinct species. Lamarck imagined that he had discovered this *vera causa* in the admitted facts that some organs may be modified by exercise ; and that modifications, once produced, are capable of hereditary transmission. It does not seem to have occurred to him to inquire whether there is

any reason to believe that there are any limits to
the amount of modification producible, or to ask
how long an animal is likely to endeavour to
gratify an impossible desire. The bird, in our
example, would surely have renounced fish dinners
long before it had produced the least effect on leg
or neck.

Since Lamarck's time, almost all competent
naturalists have left speculations on the origin of
species to such dreamers as the author of the
" Vestiges," by whose well-intentioned efforts the
Lamarckian theory received its final condemnation
in the minds of all sound thinkers. Notwithstanding this silence, however, the transmutation
theory, as it has been called, has been a " skeleton
in the closet " to many an honest zoologist and
botanist who had a soul above the mere naming of
dried plants and skins. Surely, has such an one
thought, nature is a mighty and consistent whole,
and the providential order established in the
world of life must, if we could only see it rightly
be consistent with that dominant over the multi-
form shapes of brute matter. But what is the
history of astronomy, of all the branches of physics,
of chemistry, of medicine, but a narration of the
steps by which the human mind has been compelled, often sorely against its will, to recognise
the operation of secondary causes in events where
ignorance beheld an immediate intervention of a
higher power? And when we know that living

things are formed of the same elements as the
inorganic world, that they act and react upon it,
bound by a thousand ties of natural piety, is it
probable, nay is it possible, that they, and they
alone, should have no order in their seeming
disorder, no unity in their seeming multiplicity,
should suffer no explanation by the discovery
of some central and sublime law of mutual
connection ?

Questions of this kind have assuredly often arisen,
but it might have been long before they received
such expression as would have commanded the
respect and attention of the scientific world, had
it not been for the publication of the work which
prompted this article. Its author, Mr. Darwin,
inheritor of a once celebrated name, won his spurs
in science when most of those now distinguished
were young men, and has for the last twenty
years held a place in the front ranks of British
philosophers. After a circumnavigatory voyage,
undertaken solely for the love of his science, Mr.
Darwin published a series of researches which at
once arrested the attention of naturalists and
geologists; his generalisations have since received
ample confirmation and now command universal
assent, nor is it questionable that they have had
the most important influence on the progress of
science. More recently Mr. Darwin, with a
versatility which is among the rarest of gifts,
turned his attention to a most difficult question of

zoology and minute anatomy; and no living
naturalist and anatomist has published a better
monograph than that which resulted from his
labours. Such a man, at all events, has not
entered the sanctuary with unwashed hands, and
when he lays before us the results of twenty
years' investigation and reflection we must listen
even though we be disposed to strike. But, in
reading his work, it must be confessed that the
attention which might at first be dutifully, soon
becomes willingly, given, so clear is the author's
thought, so outspoken his conviction, so honest
and fair the candid expression of his doubts.
Those who would judge the book must read it :
we shall endeavour only to make its line of argument and its philosophical position intelligible to
the general reader in our own way.

The Baker Street Bazaar has just been exhibiting its familiar annual spectacle. Straight-backed,
small-headed, big-barrelled oxen, as dissimilar
from any wild species as can well be imagined,
contended for attention and praise with sheep of
half-a-dozen different breeds and styes of bloated
preposterous pigs, no more like a wild boar or sow
than a city alderman is like an ourang-outang.
The cattle show has been, and perhaps may again
be, succeeded by a poultry show, of whose crowing
and clucking prodigies it can only be certainly
predicated that they will be very unlike the
aboriginal *Phasianus gallus*. If the seeker after

animal anomalies is not satisfied, a turn or two in
Seven Dials will convince him that the breeds of
pigeons are quite as extraordinary and unlike one
another and their parent stock, while the Horti-
cultural Society will provide him with any number
of corresponding vegetable aberrations from
nature's types.   He will learn with no little
surprise, too, in the course of his travels, that the
proprietors and producers of these animal and
vegetable anomalies regard them as distinct
species, with a firm belief, the strength of which
is exactly proportioned to their ignorance of
scientific biology, and which is the more remarkable as they are all proud of their skill in originating such "species."

On careful inquiry it is found that all these, and
the many other artificial breeds or races of animals
and plants, have been produced by one method.
The breeder—and a skilful one must be a person
of much sagacity and natural or acquired perceptive
faculty—notes some slight difference, arising he
knows not how, in some individuals of his stock.
If he wish to perpetuate the difference, to form a
breed with the peculiarity in question strongly
marked, he selects such male and female individuals as exhibit the desired character, and breeds
from them.   Their offspring are then carefully
examined, and those which exhibit the peculiarity
the most distinctly are selected for breeding; and
this operation is repeated until the desired amount

of divergence from the primitive stock is reached. It is then found that by continuing the process of selection—always breeding, that is, from well-marked forms, and allowing no impure crosses to interfere—a race may be formed, the tendency of which to reproduce itself is exceedingly strong; nor is the limit to the amount of divergence which may be thus produced known; but one thing is certain, that, if certain breeds of dogs, or of pigeons, or of horses, were known only in a fossil state, no naturalist would hesitate in regarding them as distinct species.

But in all these cases we have human interference. Without the breeder there would be no selection, and without the selection no race. Before admitting the possibility of natural species having originated in any similar way, it must be proved that there is in Nature some power which takes the place of man, and performs a selection *suâ sponte*. It is the claim of Mr. Darwin that he professes to have discovered the existence and the *modus operandi* of this "natural selection," as he terms it; and, if he be right, the process is perfectly simple and comprehensible, and irresistibly deducible from very familiar but well nigh forgotten facts.

Who, for instance, has duly reflected upon all the consequences of the marvellous struggle for existence which is daily and hourly going on among living beings? Not only does every animal live at the expense of some other animal or plant, but the very plants are at war. The ground is full of seeds that cannot rise into seedlings; the seedlings rob one another of air, light and water, the strongest robber winning the day, and extinguishing his competitors. Year after year, the wild animals with which man never interferes are, on the average, neither more nor less numerous than they were; and yet we know that the annual produce of every pair is from one to perhaps a million young; so that it is mathematically certain that, on the average, as many are killed by natural causes as are born every year, and those only escape which happen to be a little better fitted to resist destruction than those which die. The individuals of a species are like the crew of a foundered ship, and none but good swimmers have a chance of reaching the land.

Such being unquestionably the necessary conditions under which living creatures exist, Mr. Darwin discovers in them the instrument of natural selection. Suppose that in the midst of this incessant competition some individuals of a species (A) present accidental variations which happen to fit them a little better than their fellows for the struggle in which they are engaged, then the chances are in favour, not only of these individuals being better nourished than the others, but of their predominating over their fellows in other ways, and of having a better chance of leaving

offspring, which will of course tend to reproduce
the peculiarities of their parents. Their offspring
will, by a parity of reasoning, tend to predominate
over their contemporaries, and there being (suppose) no room for more than one species such as
A, the weaker variety will eventually be destroyed
by the new destructive influence which is thrown
into the scale, and the stronger will take its place.
Surrounding conditions remaining unchanged, the
new variety (which we may call B)—supposed, for
argument's sake, to be the best adapted for these
conditions which can be got out of the original
stock—will remain unchanged, all accidental devia-
tions from the type becoming at once extinguished,
as less fit for their post than B itself. The tendency of B to persist will grow with its persistence
through successive generations, and it will acquire
all the characters of a new species.

But, on the other hand, if the conditions of life
change in any degree, however slight, B may no
longer be that form which is best adapted to with-
stand their destructive, and profit by their sus-
taining, influence ; in which case if it should give
rise to a more competent variety (C), this will take
its place and become a new species ; and thus, by
natural selection, the species B and C will be successively derived from A.

That this most ingenious hypothesis enables us
to give a reason for many apparent anomalies in
the distribution of living beings in time and space,

and that it is not contradicted by the main phenomena of life and organisation appear to us to be unquestionable; and, so far, it must be admitted to have an immense advantage over any of its predecessors. But it is quite another matter to affirm absolutely either the truth or falsehood of Mr. Darwin's views at the present stage of the inquiry. Goethe has an excellent aphorism defining that state of mind which he calls " Thätige Skepsis " —active doubt. It is doubt which so loves truth that it neither dares rest in doubting, nor extinguish itself by unjustified belief; and we commend this state of mind to students of species, with respect to Mr. Darwin's or any other hypothesis, as to their origin. The combined investigations of another twenty years may, perhaps, enable naturalists to say whether the modifying causes and the selective power, which Mr. Darwin has satisfactorily shown to exist in Nature, are competent to produce all the effects he ascribes to them ; or whether, on the other hand, he has been led to over-estimate the value of the principle of natural selection, as greatly as Lamarck over-estimated his *vera causa* of modification by exercise.

But there is, at all events, one advantage possessed by the more recent writer over his predecessor. Mr. Darwin abhors mere speculation as nature abhors a vacuum. He is as greedy of cases and precedents as any constitutional lawyer, and all the principles he lays down are capable of being brought to the test of observation and experiment. The path he bids us follow professes to be, not a mere airy track, fabricated of ideal cobwebs, but a solid and broad bridge of facts. If it be so, it will carry us safely over many a chasm in our knowledge, and lead us to a region free from the snares of those fascinating but barren virgins, the Final Causes, against whom a high authority has so justly warned us. " My sons, dig in the vineyard," were the last words of the old man in the fable : and, though the sons found no treasure, they made their fortunes by the grapes.

II

THE ORIGIN OF SPECIES

[1860]

MR. DARWIN'S long-standing and well-earned scientific eminence probably renders him indifferent to that social notoriety which passes by the name of success; but if the calm spirit of the philosopher have not yet wholly superseded the ambition and the vanity of the carnal man within him, he must be well satisfied with the results of his venture in publishing the " Origin of Species." Overflowing the narrow bounds of purely scientific circles, the "species question" divides with Italy and the Volunteers the attention of general society. Everybody has read Mr. Darwin's book, or, at least, has given an opinion upon its merits or demerits; pietists, whether lay or ecclesiastic, decry it with the mild railing which sounds so charitable; bigots denounce it with ignorant invective; old ladies of both sexes consider it a

decidedly dangerous book, and even savants, who
have no better mud to throw, quote antiquated
writers to show that its author is no better than
an ape himself ; while every philosophical thinker
hails it as a veritable Whitworth gun in the
armoury of liberalism ; and all competent naturalists and physiologists, whatever their opinions as
to the ultimate fate of the doctrines put forth,
acknowledge that the work in which they are
embodied is a solid contribution to knowledge
and inaugurates a new epoch in natural history.

Nor has the discussion of the subject been
restrained within the limits of conversation.
When the public is eager and interested, reviewers
must minister to its wants ; and the genuine
littérateur is too much in the habit of acquiring
his knowledge from the book he judges—as the
Abyssinian is said to provide himself with steaks
from the ox which carries him—to be withheld
from criticism of a profound scientific work by
the mere want of the requisite preliminary scientific acquirement ; while, on the other hand, the
men of science who wish well to the new views,
no less than those who dispute their validity, have
naturally sought opportunities of expressing their
opinions.   Hence it is not surprising that almost
all the critical journals have noticed Mr. Darwin's
work at greater or less length ; and so many dis-
quisitions, of every degree of excellence, from the
poor product of ignorance, too often stimulated by

science, and having spent many years in gathering
and sifting materials for his present work, the
store of accurately registered facts upon which the
author of the " Origin of Species " is able to draw
at will is prodigious.

But this very superabundance of matter must
have been embarrassing to a writer who, for the
present, can only put forward an abstract of his
views; and thence it arises, perhaps, that notwithstanding the clearness of the style, those who
attempt fairly to digest the book find much of it
a sort of intellectual pemmican—a mass of facts
crushed and pounded into shape, rather than held
together by the ordinary medium of an obvious
logical bond; due attention will, without doubt,
discover this bond, but it is often hard to find.

Again, from sheer want of room, much has to
be taken for granted which might readily enough
be proved; and hence, while the adept, who can
supply the missing links in the evidence from his
own knowledge, discovers fresh proof of the singular thoroughness with which all difficulties have
been considered and all unjustifiable suppositions
avoided, at every reperusal of Mr. Darwin's pregnant paragraphs, the novice in biology is apt to
complain of the frequency of what he fancies is
gratuitous assumption.

Thus while it may be doubted if, for some years,
any one is likely to be competent to pronounce
judgment on all the issues raised by Mr. Darwin,

there is assuredly abundant room for him, who, assuming the humbler, though perhaps as useful, office of an interpreter between the "Origin of Species" and the public, contents himself with endeavouring to point out the nature of the problems which it discusses; to distinguish between the ascertained facts and the theoretical views which it contains; and finally, to show the extent to which the explanation it offers satisfies the requirements of scientific logic. At any rate, it is this office which we purpose to undertake in the following pages.

It may be safely assumed that our readers have a general conception of the nature of the objects to which the word "species" is applied; but it has, perhaps, occurred to a few, even to those who are naturalists *ex professo*, to reflect, that, as commonly employed, the term has a double sense and denotes two very different orders of relations. When we call a group of animals, or of plants, a species, we may imply thereby, either that all these animals or plants have some common peculiarity of form or structure; or, we may mean that they possess some common functional character. That part of biological science which deals with form and structure is called Morphology—that which concerns itself with function, Physiology—so that we may conveniently speak of these two senses, or aspects, of "species"—the one as morphological, the other as physiological. Regarded from the former point of view, a species is nothing more than a kind of animal or plant, which is distinctly definable from all others, by certain constant, and not merely sexual, morphological peculiarities. Thus horses form a species, because the group of animals to which that name is applied is distinguished from all others in the world by the following constantly associated characters. They have—1, A vertebral column ; 2, Mammæ; 3, A placental embryo ; 4, Four legs ; 5, A single well-developed toe in each foot provided with a hoof; 6, A bushy tail ; and 7, Callosities on the inner sides of both the fore and the hind legs. The asses, again, form a distinct species, because, with the same characters, as far as the fifth in the above list, all asses have tufted tails, and have callosities only on the inner side of the fore-legs. If animals were discovered having the general characters of the horse, but sometimes with callosities only on the fore-legs, and more or less tufted tails ; or animals having the general characters of the ass, but with more or less bushy tails, and sometimes with callosities on both pairs of legs, besides being intermediate in other respects—the two species would have to be merged into one. They could no longer be regarded as morphologically distinct species, for they would not be distinctly definable one from the other.

However bare and simple this definition of species may appear to be, we confidently appeal to

Fauna of the world : it is obvious that the definitions of these species can be only of a purely
structural, or morphological, character. It is
probable that naturalists would have avoided
much confusion of ideas if they had more frequently borne the necessary limitations of our
knowledge in mind. But while it may safely be
admitted that we are acquainted with only the
morphological characters of the vast majority of
species—the functional or physiological, peculiarities of a few have been carefully investigated, and
the result of that study forms a large and most
interesting portion of the physiology of reproduction.

The student of Nature wonders the more and is
astonished the less, the more conversant he becomes
with her operations; but of all the perennial
miracles she offers to his inspection, perhaps the
most worthy of admiration is the development of
a plant or of an animal from its embryo. Examine
the recently laid egg of some common animal,
such as a salamander or newt. It is a minute
spheroid in which the best microscope will reveal
nothing but a structureless sac, enclosing a glairy
fluid, holding granules in suspension.[1] But strange
possibilities lie dormant in that semi-fluid globule.
Let a moderate supply of warmth reach its watery
cradle, and the plastic matter undergoes changes

[1] [When this sentence was written, it was generally believed
that the original nucleus of the egg (the germinal vesicle)
disappeared. 1893.]

so rapid, yet so steady and purposelike in their
succession, that one can only compare them to
those operated by a skilled modeller upon a formless lump of clay. As with an invisible trowel,
the mass is divided and subdivided into smaller
and smaller portions, until it is reduced to an
aggregation of granules not too large to build withal
the finest fabrics of the nascent organism. And,
then, it is as if a delicate finger traced out the line
to be occupied by the spinal column, and moulded
the contour of the body; pinching up the head
at one end, the tail at the other, and fashioning
flank and limb into due salamandrine proportions,
in so artistic a way, that, after watching the process
hour by hour, one is almost involuntarily possessed
by the notion, that some more subtle aid to vision
than an achromatic, would show the hidden artist,
with his plan before him, striving with skilful
manipulation to perfect his work.

As life advances, and the young amphibian
ranges the waters, the terror of his insect contemporaries, not only are the nutritious particles
supplied by its prey, by the addition of which to
its frame, growth takes place, laid down, each in
its proper spot, and in such due proportion to the
rest, as to reproduce the form, the colour, and the
size, characteristic of the parental stock; but even
the wonderful powers of reproducing lost parts
possessed by these animals are controlled by the
same governing tendency. Cut off the legs, the

tail, the jaws, separately or all together, and, as Spallanzani showed long ago, these parts not only grow again, but the redintegrated limb is formed on the same type as those which were lost. The new jaw, or leg, is a newt's, and never by any accident more like that of a frog. What is true of the newt is true of every animal and of every plant; the acorn tends to build itself up again into a woodland giant such as that from whose twig it fell; the spore of the humblest lichen reproduces the green or brown incrustation which gave it birth; and at the other end of the scale of life, the child that resembled neither the paternal nor the maternal side of the house would be regarded as a kind of monster.

So that the one end to which, in all living beings, the formative impulse is tending—the one scheme which the Archæus of the old speculators strives to carry out, seems to be to mould the offspring into the likeness of the parent. It is the first great law of reproduction, that the offspring tends to resemble its parent or parents, more closely than anything else.

Science will some day show us how this law is a necessary consequence of the more general laws which govern matter; but, for the present, more can hardly be said than that it appears to be in harmony with them. We know that the phæ-nomena of vitality are not something apart from other physical phænomena, but one with them;

and matter and force are the two names of the one artist who fashions the living as well as the lifeless. Hence living bodies should obey the same great laws as other matter—nor, throughout Nature, is there a law of wider application than this, that a body impelled by two forces takes the direction of their resultant. But living bodies may be regarded as nothing but extremely complex bundles of forces held in a mass of matter, as the complex forces of a magnet are held in the steel by its coercive force; and, since the differences of sex are comparatively slight, or, in other words, the sum of the forces in each has a very similar tendency, their resultant, the offspring, may reasonably be expected to deviate but little from a course parallel to either, or to both.

Represent the reason of the law to ourselves by what physical metaphor or analogy we will, however, the great matter is to apprehend its existence and the importance of the consequences deducible from it. For things which are like to the same are like to one another; and if, in a great series of generations, every offspring is like its parent, it follows that all the offspring and all the parents must be like one another; and that, given an original parental stock, with the opportunity of undisturbed multiplication, the law in question necessitates the production, in course of time, of an indefinitely large group, the whole of the members of which are at once very similar and are blood relations, having descended from the same parent, or pair of parents. The proof that all the members of any given group of animals, or plants, had thus descended, would be ordinarily considered sufficient to entitle them to the rank of physiological species, for most physiologists consider species to be definable as "the offspring of a single primitive stock."

But though it is quite true that all those groups we call species *may*, according to the known laws of reproduction, have descended from a single stock, and though it is very likely they really have done so, yet this conclusion rests on deduction and can hardly hope to establish itself upon a basis of observation. And the primitiveness of the supposed single stock, which, after all, is the essential part of the matter, is not only a hypothesis, but one which has not a shadow of foundation, if by " primitive " be meant "independent of any other living being." A scientific definition, of which an unwarrantable hypothesis forms an essential part, carries its condemnation within itself; but, even supposing such a definition were, in form, tenable, the physiologist who should attempt to apply it in Nature would soon find himself involved in great, if not inextricable, difficulties. As we have said, it is indubitable that offspring *tend* to resemble the parental organism, but it is equally true that the similarity attained never amounts to identity either in form or in structure. There is always a certain amount of deviation, not only from the precise characters of a single parent, but when, as in most animals and many plants, the sexes are lodged in distinct individuals, from an exact mean between the two parents. And indeed, on general principles, this slight deviation seems as intelligible as the general similarity, if we reflect how complex the co-operating "bundles of forces" are, and how improbable it is that, in any case, their true resultant shall coincide with any mean between the more obvious characters of the two parents. Whatever be its cause, however, the co-existence of this tendency to minor variation with the tendency to general similarity, is of vast importance in its bearing on the question of the origin of species.

As a general rule, the extent to which an offspring differs from its parent is slight enough; but, occasionally, the amount of difference is much more strongly marked, and then the divergent offspring receives the name of a Variety. Multitudes, of what there is every reason to believe are such varieties, are known, but the origin of very few has been accurately recorded, and of these we will select two as more especially illustrative of the main features of variation. The first of them is that of the "Ancon" or "Otter" sheep, of which a careful account is given by Colonel David Humphreys, F.R.S., in a letter to Sir

Joseph Banks, published in the "Philosophical Transactions" for 1813. It appears that one Seth Wright, the proprietor of a farm on the banks of the Charles River, in Massachusetts, possessed a flock of fifteen ewes and a ram of the ordinary kind. In the year 1791, one of the ewes presented her owner with a male lamb, differing, for no assignable reason, from its parents by a proportionally long body and short bandy legs, whence it was unable to emulate its relatives in those sportive leaps over the neighbours' fences, in which they were in the habit of indulging, much to the good farmer's vexation.

The second case is that detailed by a no less unexceptionable authority than Réaumur, in his "Art de faire éclore les Poulets." A Maltese couple, named Kelleia, whose hands and feet were constructed upon the ordinary human model, had born to them a son, Gratio, who possessed six perfectly movable fingers on each hand, and six toes, not quite so well formed, on each foot. No cause could be assigned for the appearance of this unusual variety of the human species.

Two circumstances are well worthy of remark in both these cases. In each, the variety appears to have arisen in full force, and, as it were, *per saltum* ; a wide and definite difference appearing, at once, between the Ancon ram and the ordinary sheep ; between the six-fingered and six-toed Gratio Kelleia and ordinary men. In neither case is it possible to point out any obvious reason for the appearance of the variety. Doubtless there were determining causes for these as for all other phænomena; but they do not appear, and we can be tolerably certain that what are ordinarily understood as changes in physical conditions, as in climate, in food, or the like, did not take place and had nothing to do with the matter. It was no case of what is commonly called adaptation to circumstances; but, to use a conveniently erroneous phrase, the variations arose spontaneously. The fruitless search after final causes leads their pursuers a long way; but even those hardy teleologists, who are ready to break through all the laws of physics in chase of their favourite will-o'-the-wisp, may be puzzled to discover what purpose could be attained by the stunted legs of Seth Wright's ram or the hexadactyle members of Gratio Kelleia.

Varieties then arise we know not why; and it is more than probable that the majority of varieties have arisen in this "spontaneous" manner, though we are, of course, far from denying that they may be traced, in some cases, to distinct external influences; which are assuredly competent to alter the character of the tegumentary covering, to change colour, to increase or diminish the size of muscles, to modify constitution, and, among plants, to give rise to the metamorphosis of stamens into petals, and so forth. But however they may have arisen, what especially interests us at present is, to

remark that, once in existence, many varieties obey
the fundamental law of reproduction that like tends
to produce like; and their offspring exemplify it by
tending to exhibit the same deviation from the
parental stock as themselves. Indeed, there seems
to be, in many instances, a prepotent influence
about a newly-arisen variety which gives it what
one may call an unfair advantage over the normal
descendants from the same stock. This is strik-
ingly exemplified by the case of Gratio Kelleia,
who married a woman with the ordinary pentadactyle extremities, and had by her four children,
Salvator, George, André, and Marie. Of these
children Salvator, the eldest boy, had six fingers
and six toes, like his father ; the second and third,
also boys, had five fingers and five toes, like their
mother, though the hands and feet of George
were slightly deformed. The last, a girl, had five
fingers and five toes, but the thumbs were slightly
deformed. The variety thus reproduced itself
purely in the eldest, while the normal type
reproduced itself purely in the third, and almost
purely in the second and last : so that it would
seem, at first, as if the normal type were more
powerful than the variety. But all these children
grew up and intermarried with normal wives and
husband, and then, note what took place : Salvator
had four children, three of whom exhibited the
hexadactyle members of their grandfather and
father, while the youngest had the pentadactyle

enforced by Nature upon the newly-arrived ram; and they advised Wright to kill the old patriarch of his fold, and install the Ancon ram in his place. The result justified their sagacious anticipations, and coincided very nearly with what occurred to the progeny of Gratio Kelleia. The young lambs were almost always either pure Ancons, or pure ordinary sheep.[1] But when sufficient Ancon sheep were obtained to interbreed with one another, it was found that the offspring was always pure Ancon. Colonel Humphreys, in fact, states that he was acquainted with only "one questionable case of a contrary nature." Here, then, is a remarkable and well-established instance, not only of a very distinct race being established *per saltum*, but of that race breeding "true" at once, and showing no mixed forms, even when crossed with another breed.

By taking care to select Ancons of both sexes, for breeding from, it thus became easy to establish an extremely well-marked race; so peculiar that,

---

[1] Colonel Humphreys' statements are exceedingly explicit on this point :—"When an Ancon ewe is impregnated by a common ram, the increase resembles wholly either the ewe or the ram. The increase of the common ewe impregnated by an Ancon ram follows entirely the one or the other, without blending any of the distinguishing and essential peculiarities of both. Frequent instances have happened where common ewes have had twins by Ancon rams, when one exhibited the complete marks and features of the ewe, the other of the ram. The contrast has been rendered singularly striking, when one short-legged and one long-legged lamb, produced at a birth, have been seen sucking the dam at the same time."—*Philosophical Transactions*, 1813, Pt. I. pp. 89, 90.

even when herded with other sheep, it was noted that the Ancons kept together. And there is every reason to believe that the existence of this breed might have been indefinitely protracted; but the introduction of the Merino sheep, which were not only very superior to the Ancons in wool and meat, but quite as quiet and orderly, led to the complete neglect of the new breed, so that, in 1813, Colonel Humphreys found it difficult to obtain the specimen, the skeleton of which was presented to Sir Joseph Banks. We believe that, for many years, no remnant of it has existed in the United States.

Gratio Kelleia was not the progenitor of a race of six-fingered men, as Seth Wright's ram became a nation of Ancon sheep, though the tendency of the variety to perpetuate itself appears to have been fully as strong in the one case as in the other. And the reason of the difference is not far to seek. Seth Wright took care not to weaken the Ancon blood by matching his Ancon ewes with any but males of the same variety, while Gratio Kelleia's sons were too far removed from the patriarchal times to intermarry with their sisters; and his grand-children seem not to have been attracted by their six-fingered cousins. In other words, in the one example a race was produced, because, for several generations, care was taken to *select* both parents of the breeding stock from animals exhibiting a tendency to vary in the

same direction ; while, in the other, no race was
evolved, because no such selection was exercised.
A race is a propagated variety ; and as, by the laws
of reproduction, offspring tend to assume the
parental forms, they will be more likely to propagate a variation exhibited by both parents than
that possessed by only one.

There is no organ of the body of an animal
which may not, and does not, occasionally, vary
more or less from the normal type ; and there is no
variation which may not be transmitted and which,
if selectively transmitted, may not become the
foundation of a race.   This great truth, sometimes
forgotten by philosophers, has long been familiar
to practical agriculturists and breeders ; and upon
it rest all the methods of improving the breeds of
domestic animals, which, for the last century, have
been followed with so much success in England.
Colour, form, size, texture of hair or wool, proportions of various parts, strength or weakness of
constitution, tendency to fatten or to remain lean,
to give much or little milk, speed, strength, tem-
per, intelligence, special instincts ; there is not one
of these characters the transmission of which is not
an every-day occurrence within the experience of
cattle-breeders, stock-farmers, horse-dealers, and
dog and poultry fanciers.   Nay, it is only the other
day that an eminent physiologist, Dr. Brown-
Séquard, communicated to the Royal Society his
discovery that epilepsy, artificially produced in

guinea-pigs, by a means which he has discovered, is transmitted to their offspring.[1]

But a race, once produced, is no more a fixed and immutable entity than the stock whence it sprang ; variations arise among its members, and as these variations are transmitted like any others, new races may be developed out of the pre-existing one *ad infinitum*, or, at least, within any limit at present determined. Given sufficient time and sufficiently careful selection, and the multitude of races which may arise from a common stock is as astonishing as are the extreme structural differences which they may present. A remarkable example of this is to be found in the rock-pigeon, which Mr. Darwin has, in our opinion, satisfactorily demonstrated to be the progenitor of all our domestic pigeons, of which there are certainly more than a hundred well-marked races. The most noteworthy of these races are, the four great stocks known to the " fancy " as tumblers, pouters, carriers, and fantails; birds which not only differ most singularly in size, colour, and habits, but in the form of the beak and of the skull : in the proportions of the beak to the skull; in the number of tail-feathers ; in the absolute and relative size of the feet ; in the presence or absence of the uropygial gland ; in the number of vertebræ in the back ; in short, in precisely those characters in which

[1] [Compare Weismann's *Essays Upon Heredity*, p. 310, *et seq.* 1893.]

the genera and species of birds differ from one another.

And it is most remarkable and instructive to observe, that none of these races can be shown to have been originated by the action of changes in what are commonly called external circumstances, upon the wild rock-pigeon. On the contrary, from time immemorial pigeon-fanciers have had essentially similar methods of treating their pets, which have been housed, fed, protected and cared for in much the same way in all pigeonries. In fact, there is no case better adapted than that of the pigeons to refute the doctrine which one sees put forth on high authority, that "no other characters than those founded on the development of bone for the attachment of muscles" are capable of variation. In precise contradiction of this hasty assertion, Mr. Darwin's researches prove that the skeleton of the wings in domestic pigeons has hardly varied at all from that of the wild type ; while, on the other hand, it is in exactly those respects, such as the relative length of the beak and skull, the number of the vertebræ, and the number of the tail-feathers, in which muscular exertion can have no important influence, that the utmost amount of variation has taken place.

We have said that the following out of the properties exhibited by physiological species would lead us into difficulties, and at this point they begin

to be obvious; for if, as the result of spontaneou
variation and of selective breeding, the progeny of
a common stock may become separated into groups
distinguished from one another by constant, not
sexual, morphological characters, it is clear that
the physiological definition of species is likely to
clash with the morphological definition. No one
would hesitate to describe the pouter and the
tumbler as distinct species, if they were found fossil,
or if their skins and skeletons were imported, as
those of exotic wild birds commonly are—and without doubt, if considered alone, they are good and
distinct morphological species. On the other hand,
they are not physiological species, for they are
descended from a common stock, the rock-pigeon.

Under these circumstances, as it is admitted on
all sides that races occur in Nature, how are we to
know whether any apparently distinct animals are
really of different physiological species, or not,
seeing that the amount of morphological difference
is no safe guide? Is there any test of a physiological species? The usual answer of physiologists
is in the affirmative. It is said that such a test is
to be found in the phænomena of hybridisation—
in the results of crossing races, as compared with
the results of crossing species.

So far as the evidence goes at present, individuals, of what are certainly known to be mere
races produced by selection, however distinct they
may appear to be, not only breed freely together,

but the offspring of such crossed races are perfectly fertile with one another. Thus, the spaniel and the greyhound, the dray-horse and the Arab, the pouter and the tumbler, breed together with perfect freedom, and their mongrels, if matched with other mongrels of the same kind, are equally fertile.

On the other hand, there can be no doubt that the individuals of many natural species are either absolutely infertile if crossed with individuals of other species, or, if they give rise to hybrid offspring, the hybrids so produced are infertile when paired together. The horse and the ass, for instance, if so crossed, give rise to the mule, and there is no certain evidence of offspring ever having been produced by a male and female mule. The unions of the rock-pigeon and the ring-pigeon appear to be equally barren of result. Here, then, says the physiologist, we have a means of distinguishing any two true species from any two varieties. If a male and a female, selected from each group, produce offspring, and that offspring is fertile with others produced in the same way, the groups are races and not species. If, on the other hand, no result ensues, or if the offspring are infertile with others produced in the same way, they are true physiological species. The test would be an admirable one, if, in the first place, it were always practicable to apply it, and if, in the second, it always yielded results susceptible of a definite interpretation. Unfortunately, species are ineffective with the females of the first. So that, in the last-named instance, a physiologist, who should cross the two species in one way, would decide that they were true species; while another, who should cross them in the reverse way, would, with equal justice, according to the rule, pronounce them to be mere races. Several plants, which there is great reason to believe are mere varieties, are almost sterile when crossed; while both animals and plants, which have always been regarded by naturalists as of distinct species, turn out, when the test is applied, to be perfectly fertile. Again, the sterility or fertility of crosses seems to bear no relation to the structural resemblances or differences of the members of any two groups.

Mr. Darwin has discussed this question with singular ability and circumspection, and his conclusions are summed up as follows, at page 276 of his work :—

" First crosses between forms sufficiently distinct to be ranked as species, and their hybrids, are very generally, but not universally, sterile. The sterility is of all degrees, and is often so slight that the two most careful experimentalists who have ever lived have come to diametrically opposite conclusions in ranking forms by this test. The sterility is innately variable in individuals of the same species, and is eminently susceptible of favourable and unfavourable conditions. The degree of sterility does not strictly follow systematic affinity, but is governed by several curious and complex laws. It is generally different and sometimes widely different, in reciprocal crosses between the same two species. It is not always equal in degree in a first cross, and in the hybrid produced from this cross.

" In the same manner as in grafting trees, the capacity of one species or variety to take on another is incidental on generally unknown differences in their vegetative systems ; so in crossing, the greater or less facility of one species to unite with another is incidental on unknown differences in their reproductive systems. There is no more reason to think that species have been specially endowed with various degrees of sterility to prevent them crossing and breeding in Nature, than to think that trees have been specially endowed with various and somewhat analogous degrees of difficulty in being grafted together, in order to prevent them becoming inarched in our forests.

" The sterility of first crosses between pure species, which have their reproductive systems perfect, seems to depend on several circumstances ; in some cases largely on the early death of the embryo. The sterility of hybrids which have their reproductive systems imperfect, and which have had this system and their whole organisation disturbed by being compounded of two distinct species, seems closely allied to that sterility which so frequently affects pure species when their natural conditions of life have been disturbed. This view is supported by a parallelism of another kind : namely, that the crossing of forms, only slightly different, is favourable to the vigour and fertility of the offspring ; and that slight changes in the conditions of life are apparently favourable to the vigour and fertility of all organic beings. It is not surprising that the degree of difficulty in uniting two species, and the degree of sterility of their hybrid offspring, should generally correspond, though due to distinct causes ; for both depend on the amount of difference of some kind between the species which are crossed. Nor is it surprising that the facility of effecting a first cross, the fertility of hybrids produced from it, and the capacity of being grafted together—though this latter capacity evidently depends on widely different circumstances—should all run to a certain extent parallel with the systematic affinity of the forms which are subjected to experiment ; for systematic affinity attempts to express all kinds of resemblance between all species.

"First crosses between forms known to be varieties, or sufficiently alike to be considered as varieties, and their mongrel offspring, are very generally, but not quite universally, fertile. Nor is this nearly general and perfect fertility surprising, when we remember how liable we are to argue in a circle with respect to varieties in a state of Nature ; and when we remember that the greater number of varieties have been produced under domestication by the selection of mere external differences, and not of differences in the reproductive system. In all other respects, excluding fertility, there is a close general resemblance between hybrids and mongrels."—Pp. 276—8.

We fully agree with the general tenor of this weighty passage ; but forcible as are these arguments, and little as the value of fertility or infertility as a test of species may be, it must not be forgotten that the really important fact, so far as the inquiry into the origin of species goes, is, that there are such things in Nature as groups of animals and of plants, the members of which are incapable of fertile union with those of other groups ; and that there are such things as hybrids, which are absolutely sterile when crossed with other hybrids. For, if such phænomena as these were exhibited by only two of those assemblages of living objects, to which the name of species (whether it be used in its physiological or in its morphological sense) is given, it would have to be accounted for by any theory of the origin of species, and every theory which could not account for it would be, so far, imperfect.

Up to this point, we have been dealing with matters of fact, and the statements which we have laid before the reader would, to the best of our knowledge, be admitted to contain a fair exposition of what is at present known respecting the essential properties of species, by all who have studied the question. And whatever may be his theoretical views, no naturalist will probably be disposed to demur to the following summary of that exposition :—

Living beings, whether animals or plants, are divisible into multitudes of distinctly definable kinds, which are morphological species. They are also divisible into groups of individuals, which breed freely together, tending to reproduce their like, and are physiological species. Normally resembling their parents, the offspring of members of these species are still liable to vary ; and the variation may be perpetuated by selection, as a race, which race, in many cases, presents all the characteristics of a morphological species. But it is not as yet proved that a race ever exhibits, when crossed with another race of the same species, those phænomena of hybridisation which are exhibited by many species when crossed with other species. On the other hand, not only is it not proved that all species give rise to hybrids infertile *inter se*, but there is much reason to believe that, in crossing, species exhibit every gradation from perfect sterility to perfect fertility.

Such are the most essential characteristics of species. Even were man not one of them—a member of the same system and subject to the same laws—the question of their origin, their causal connexion, that is, with the other phænomena of the universe, must have attracted his attention, as soon as his intelligence had raised itself above the level of his daily wants.

Indeed history relates that such was the case, and has embalmed for us the speculations upon the origin of living beings, which were among the earliest products of the dawning intellectual activity of man. In those early days positive knowledge was not to be had, but the craving after it needed, at all hazards, to be satisfied, and according to the country, or the turn of thought, of the speculator, the suggestion that all living things arose from the mud of the Nile, from a primeval egg, or from some more anthropomorphic agency, afforded a sufficient resting-place for his curiosity. The myths of Paganism are as dead as Osiris or Zeus, and the man who should revive them, in opposition to the knowledge of our time, would be justly laughed to scorn ; but the coeval imaginations current among the rude inhabitants of Palestine, recorded by writers whose very name and age are admitted by every scholar to be unknown, have unfortunately not yet shared their fate, but, even at this day, are regarded by nine-tenths of the civilised world as the authoritative standard of fact and the criterion of the justice of scientific conclusions, in all that relates to the origin of things, and, among them, of species. In this nineteenth century, as at the dawn of modern physical science, the cosmogony of the semi-barbarous Hebrew is the incubus of the philosopher and the opprobrium of the orthodox. Who shall number the patient and earnest seekers after truth, from the days of Galileo until now, whose lives have been embittered and their good name blasted by the mistaken zeal of Bibliolaters? Who shall count the host of weaker men whose sense of truth has been destroyed in the effort to harmonise impossibilities—whose life has been wasted in the attempt to force the generous new wine of Science into the old bottles of Judaism, compelled by the outcry of the same strong party?

It is true that if philosophers have suffered, their cause has been amply avenged. Extinguished theologians lie about the cradle of every science as the strangled snakes beside that of Hercules; and history records that whenever science and orthodoxy have been fairly opposed, the latter has been forced to retire from the lists, bleeding and crushed if not annihilated; scotched, if not slain. But orthodoxy is the Bourbon of the world of thought. It learns not, neither can it forget; and though, at present, bewildered and afraid to move, it is as willing as ever to insist that the first chapter of Genesis contains the beginning and the end of sound science; and to visit, with such petty thunderbolts as its half-paralysed hands can hurl, those who refuse to degrade Nature to the level of primitive Judaism.

Philosophers, on the other hand, have no such aggressive tendencies. With eyes fixed on the noble goal to which "per aspera et ardua" they tend, they may, now and then, be stirred to momentary wrath by the unnecessary obstacles with which the ignorant, or the malicious, encumber, if they cannot bar, the difficult path; but why should their souls be deeply vexed? The majesty of Fact is on their side, and the elemental forces of Nature are working for them. Not a star comes to the meridian at its calculated time but testifies to the justice of their methods—their beliefs are "one with the falling rain and with the growing corn." By doubt they are established, and open inquiry is their bosom friend. Such men have no fear of traditions however venerable, and no respect for them when they become mischievous and obstructive; but they have better than mere antiquarian business in hand, and if dogmas, which ought to be fossil but are not, are not forced upon their notice, they are too happy to treat them as non-existent.

The hypotheses respecting the origin of species which profess to stand upon a scientific basis, and, as such, alone demand serious attention, are of two kinds. The one, the "special creation" hypothesis, presumes every species to have originated from one or more stocks, these not being the result of the modification of any other form of living matter—or arising by natural agencies—but being produced, as such, by a supernatural creative act.

The other, the so-called "transmutation" hypothesis, considers that all existing species are the result of the modification of pre-existing species, and those of their predecessors, by agencies similar to those which at the present day produce varieties and races, and therefore in an altogether natural way; and it is a probable, though not a necessary consequence of this hypothesis, that all living beings have arisen from a single stock. With respect to the origin of this primitive stock, or stocks, the doctrine of the origin of species is obviously not necessarily concerned. The transmutation hypothesis, for example, is perfectly consistent either with the conception of a special creation of the primitive germ, or with the supposition of its having arisen, as a modification of inorganic matter, by natural causes.

The doctrine of special creation owes its existence very largely to the supposed necessity of making science accord with the Hebrew cosmogony; but it is curious to observe that, as the doctrine is at present maintained by men of science, it is as hopelessly inconsistent with the Hebrew view as any other hypothesis.

If there be any result which has come more

ages. The other formations not uncommonly exhibit 60, 80, or even 94 per cent. of genera in common with those whose remains are imbedded in their predecessor. Not only is this true, but the subdivisions of each formation exhibit new species characteristic of, and found only in, them ; and, in many cases, as in the lias for example, the separate beds of these subdivisions are distinguished by well-marked and peculiar forms of life. A section, a hundred feet thick, will exhibit, at different heights, a dozen species of ammonite, none of which passes beyond its particular zone of limestone, or clay, into the zone below it or into that above it ; so that those who adopt the doctrine of special creation must be prepared to admit, that at intervals of time, corresponding with the thickness of these beds, the Creator thought fit to interfere with the natural course of events for the purpose of making a new ammonite. It is not easy to transplant oneself into the frame of mind of those who can accept such a conclusion as this, on any evidence short of absolute demonstration ; and it is difficult to see what is to be gained by so doing, since, as we have said, it is obvious that such a view of the origin of living beings is utterly opposed to the Hebrew cosmogony. Deserving no aid from the powerful arm of Bibliolatry, then, does the received form of the hypothesis of special creation derive any support from science or sound logic ? Assuredly

Or, lastly, let us ask ourselves whether any amount of evidence which the nature of our faculties permits us to attain, can justify us in asserting that any phænomenon is out of the reach of natural causation. To this end it is obviously necessary that we should know all the consequences to which all possible combinations, continued through unlimited time, can give rise. If we knew these, and found none competent to originate species, we should have good ground for denying their origin by natural causation. Till we know them, any hypothesis is better than one which involves us in such miserable presumption.

But the hypothesis of special creation is not only a mere specious mask for our ignorance; its existence in Biology marks the youth and imperfection of the science. For what is the history of every science but the history of the elimination of the notion of creative, or other interferences, with the natural order of the phænomena which are the subject-matter of that science? When Astronomy was young "the morning stars sang together for joy," and the planets were guided in their courses by celestial hands. Now, the harmony of the stars has resolved itself into gravitation according to the inverse squares of the distances, and the orbits of the planets are deducible from the laws of the forces which allow a schoolboy's stone to break a window. The lightning was the angel of the Lord; but it has pleased

Providence, in these modern times, that science should make it the humble messenger of man, and we know that every flash that shimmers about the horizon on a summer's evening is determined by ascertainable conditions, and that its direction and brightness might, if our knowledge of these were great enough, have been calculated.

The solvency of great mercantile companies rests on the validity of the laws which have been ascertained to govern the seeming irregularity of that human life which the moralist bewails as the most uncertain of things ; plague, pestilence, and famine are admitted, by all but fools, to be the natural result of causes for the most part fully within human control, and not the unavoidable tortures inflicted by wrathful Omnipotence upon His helpless handiwork.

Harmonious order governing eternally continuous progress—the web and woof of matter and force interweaving by slow degrees, without a broken thread, that veil which lies between us and the Infinite—that universe which alone we know or can know ; such is the picture which science draws of the world, and in proportion as any part of that picture is in unison with the rest, so may we feel sure that it is rightly painted. Shall Biology alone remain out of harmony with her sister sciences ?

Such arguments against the hypothesis of the direct creation of species as these are plainly

enough deducible from general considerations; but
there are, in addition, phænomena exhibited by
species themselves, and yet not so much a part of
their very essence as to have required earlier
mention, which are in the highest degree per-
plexing, if we adopt the popularly accepted
hypothesis. Such are the facts of distribution in
space and in time; the singular phænomena
brought to light by the study of development;
the structural relations of species upon which our
systems of classification are founded; the great
doctrines of philosophical anatomy, such as
that of homology, or of the community of
structural plan exhibited by large groups of
species differing very widely in their habits and
functions.

The species of animals which inhabit the sea on
opposite sides of the isthmus of Panama are
wholly distinct;[1] the animals and plants which
inhabit islands are commonly distinct from those
of the neighbouring mainlands, and yet have a
similarity of aspect. The mammals of the latest
tertiary epoch in the Old and New Worlds belong
to the same genera, or family groups, as those
which now inhabit the same great geographical
area. The crocodilian reptiles which existed in the
earliest secondary epoch were similar in general
structure to those now living, but exhibit slight

---

[1] Recent investigations tend to show that this statement is
not strictly accurate.—1870.

differences in their vertebræ, nasal passages, and
one or two other points. The guinea-pig has
teeth which are shed before it is born, and hence
can never subserve the masticatory purpose for
which they seem contrived, and, in like manner,
the female dugong has tusks which never cut the
gum. All the members of the same great group
run through similar conditions in their development, and all their parts, in the adult state, are
arranged according to the same plan. Man is
more like a gorilla than a gorilla is like a lemur.
Such are a few, taken at random, among the
multitudes of similar facts which modern research
has established ; but when the student seeks for
an explanation of them from the supporters of
the received hypothesis of the origin of species,
the reply he receives is, in substance, of Oriental
simplicity and brevity—" Mashallah ! it so pleases
God !" There are different species on opposite
sides of the isthmus of Panama, because they were
created different on the two sides. The pliocene
mammals are like the existing ones, because such
was the plan of creation ; and we find rudimental
organs and similarity of plan, because it has
pleased the Creator to set before Himself a
" divine exemplar or archetype," and to copy it in
His works ; and somewhat ill, those who hold this
view imply, in some of them. That such verbal
hocus-pocus should be received as science will one
day be regarded as evidence of the low state of

General for Egypt kept his theories to himself throughout a long life, for "Telliamed," the only scientific work which is known to have proceeded from his pen, was not printed till 1735, when its author had reached the ripe age of seventy-nine; and though De Maillet lived three years longer, his book was not given to the world before 1748. Even then it was anonymous to those who were not in the secret of the anagrammatic character of its title; and the preface and dedication are so worded as, in case of necessity, to give the printer a fair chance of falling back on the excuse that the work was intended for a mere *jeu d'esprit*.

The speculations of the suppositious Indian sage, though quite as sound as those of many a "Mosaic Geology," which sells exceedingly well, have no great value if we consider them by the light of modern science. The waters are supposed to have originally covered the whole globe; to have deposited the rocky masses which compose its mountains by processes comparable to those which are now forming mud, sand, and shingle; and then to have gradually lowered their level, leaving the spoils of their animal and vegetable inhabitants embedded in the strata. As the dry land appeared, certain of the aquatic animals are supposed to have taken to it, and to have become gradually adapted to terrestrial and aërial modes of existence. But if we regard the general tenor and style of the reasoning in relation to the state of knowledge of the day, two circumstances appear very well worthy of remark. The first, that De Maillet had a notion of the modifiability of living forms (though without any precise information on the subject), and how such modifiability might account for the origin of species; the second, that he very clearly apprehended the great modern geological doctrine, so strongly insisted upon by Hutton, and so ably and comprehensively expounded by Lyell, that we must look to existing causes for the explanation of past geological events. Indeed, the following passage of the preface, in which De Maillet is supposed to speak of the Indian philosopher Telliamed, his *alter ego*, might have been written by the most philosophical uniformitarian of the present day:—

"Ce qu'il y a d'étonnant, est que pour arriver à ces connoissances il semble avoir perverti l'ordre naturel, puisqu'au lieu de s'attacher d'abord à rechercher l'origine de notre globe il a commencé par travailler à s'instruire de la nature. Mais à l'entendre, ce renversement de l'ordre a été pour lui l'effet d'un génie favorable qui l'a conduit pas à pas et comme par la main aux découvertes les plus sublimes. C'est en décomposant la substance de ce globe par une anatomie exacte de toutes ses parties qu'il a premièrement appris de quelles matières il était composé et quels arrangemens ces mêmes matières observaient entre elles. Ces lumières jointes à l'esprit de comparaison toujours nécessaire à quiconque entreprend de percer les voiles dont la nature aime à se cacher, ont servi de guide à notre philosophe pour parvenir à des connoissances plus intéressantes. Par la matière et l'arrangement de ces compositions il prétend avoir reconnu quelle est la véritable origine de ce globe que nous habitons, comment et par qui il a été formé."—Pp. xix. xx.

But De Maillet was before his age, and as could hardly fail to happen to one who speculated on a zoological and botanical question before Linnæus, and on a physiological problem before Haller, he fell into great errors here and there ; and hence, perhaps, the general neglect of his work. Robinet's speculations are rather behind, than in advance of, those of De Maillet ; and though Linnæus may have played with the hypothesis of transmutation, it obtained no serious support until Lamarck adopted it, and advocated it with great ability in his " Philosophie Zoologique."

Impelled towards the hypothesis of the transmutation of species, partly by his general cosmological and geological views ; partly by the conception of a graduated, though irregularly branching, scale of being, which had arisen out of his profound study of plants and of the lower forms of animal life, Lamarck, whose general line of thought often closely resembles that of De Maillet, made a great advance upon the crude and merely speculative manner in which that writer deals with the question of the origin of living beings, by endeavouring to find physical causes competent to effect that change of one species into another, which De Maillet had only supposed to occur. And Lamarck conceived that he had found in Nature such causes, amply sufficient for the purpose in view. It is a physiological fact, he says, that organs are increased in size by action, atrophied by inaction; it is another physiological fact that modifications produced are transmissible to offspring. Change the actions of an animal, therefore, and you will change its structure, by increasing the development of the parts newly brought into use and by the diminution of those less used; but by altering the circumstances which surround it you will alter its actions, and hence, in the long run, change of circumstance must produce change of organisation. All the species of animals, therefore, are, in Lamarck's view, the result of the indirect action of changes of circumstanc , upon those primitive germs which he considered to have originally arisen, by spontaneous generation, within the waters of the globe. It is curious, however, that Lamarck should insist so strongly [1] as he has done, that circumstances never in any degree directly modify the form or the organisation of animals, but only operate by changing their wants and consequently their actions; for he thereby brings upon himself the obvious question, How, then, do plants, which cannot be said to have wants or actions, become modified? To this he replies, that they are modified by the changes in their nutritive processes, which are effected by changing circumstances; and it does not seem to have

---

[1] See *Phil. Zoologique*, vol. i. p. 222, et seq.

occurred to him that such changes might be as well supposed to take place among animals.

When we have said that Lamarck felt that mere speculation was not the way to arrive at the origin of species, but that it was necessary, in order to the establishment of any sound theory on the subject, to discover by observation or otherwise, some *vera causa*, competent to give rise to them; that he affirmed the true order of classification to coincide with the order of their development one from another; that he insisted on the necessity of allowing sufficient time, very strongly; and that all the varieties of instinct and reason were traced back by him to the same cause as that which has given rise to species, we have enumerated his chief contributions to the advance of the question. On the other hand, from his ignorance of any power in Nature competent to modify the structure of animals, except the development of parts, or atrophy of them, in consequence of a change of needs, Lamarck was led to attach infinitely greater weight than it deserves to this agency, and the absurdities into which he was led have met with deserved condemnation. Of the struggle for existence, on which, as we shall see, Mr. Darwin lays such great stress, he had no conception; indeed, he doubts whether there really are such things as extinct species, unless they be such large animals as may have met their death at the

hands of man ; and so little does he dream of
there being any other destructive causes at work,
that, in discussing the possible existence of fossil
shells, he asks, "Pourquoi d'ailleurs seroient-ils
perdues dès que l'homme n'a pu opérer leur
destruction ? " (" Phil. Zool.," vol. i. p. 77.)   Of
the influence of selection Lamarck has as little
notion, and he makes no use of the wonderful
phænomena which are exhibited by domesticated
animals, and illustrate its powers.   The vast
influence of Cuvier was employed against the
Lamarckian views, and, as the untenability of
some of his conclusions was easily shown, his
doctrines sank under the opprobrium of scientific,
as well as of theological, heterodoxy.   Nor have
the efforts made of late years to revive them
tended to re-establish their credit in the minds of
sound thinkers acquainted with the facts of the
case ; indeed it may be doubted whether Lamarck
has not suffered more from his friends than from
his foes.

Two years ago, in fact, though we venture to
question if even the strongest supporters of the
special creation hypothesis had not, now and then,
an uneasy consciousness that all was not right,
their position seemed more impregnable than ever,
if not by its own inherent strength, at any rate by
the obvious failure of all the attempts which had
been made to carry it.   On the other hand, however much the few, who thought deeply on the

question of species, might be repelled by the generally received dogmas, they saw no way of escaping from them save by the adoption of suppositions so little justified by experiment or by observation as to be at least equally distasteful.

The choice lay between two absurdities and a middle condition of uneasy scepticism; which last, however unpleasant and unsatisfactory, was obviously the only justifiable state of mind under the circumstances.

Such being the general ferment in the minds of naturalists, it is no wonder that they mustered strong in the rooms of the Linnæan Society, on the 1st of July of the year 1858, to hear two papers by authors living on opposite sides of the globe, working out their results independently, and yet professing to have discovered one and the same solution of all the problems connected with species. The one of these authors was an able naturalist, Mr. Wallace, who had been employed for some years in studying the productions of the islands of the Indian Archipelago, and who had forwarded a memoir embodying his views to Mr. Darwin, for communication to the Linnæan Society. On perusing the essay, Mr. Darwin was not a little surprised to find that it embodied some of the leading ideas of a great work which he had been preparing for twenty years, and parts of which, containing a development of the very same views, had been perused by his private friends fifteen or sixteen years before. Perplexed in what manner to do full justice both to his friend and to himself, Mr. Darwin placed the matter in the hands of Dr. Hooker and Sir Charles Lyell, by whose advice he communicated a brief abstract of his own views to the Linnæan Society, at the same time that Mr. Wallace's paper was read. Of that abstract, the work on the " Origin of Species " is an enlargement; but a complete statement of Mr. Darwin's doctrine is looked for in the large and well-illustrated work which he is said to be preparing for publication.

The Darwinian hypothesis has the merit of being eminently simple and comprehensible in principle, and its essential positions may be stated in a very few words : all species have been produced by the development of varieties from common stocks ; by the conversion of these, first into permanent races and then into new species, by the process of *natural selection*, which process is essentially identical with that artificial selection by which man has originated the races of domestic animals—the *struggle for existence* taking the place of man, and exerting, in the case of natural selection, that selective action which he performs in artificial selection.

The evidence brought forward by Mr. Darwin in support of his hypothesis is of three kinds. First,

he endeavours to prove that species may be
originated by selection ; secondly, he attempts to
show that natural causes are competent to exert
selection ; and thirdly, he tries to prove that the
most remarkable and apparently anomalous
phænomena exhibited by the distribution,
development, and mutual relations of species,
can be shown to be deducible from the general
doctrine of their origin, which he propounds,
combined with the known facts of geological
change ; and that, even if all these phænomena
are not at present explicable by it, none are
necessarily inconsistent with it.

There cannot be a doubt that the method of
inquiry which Mr. Darwin has adopted is not only
rigorously in accordance with the canons of
scientific logic, but that it is the only adequate
method. Critics exclusively trained in classics or
in mathematics, who have never determined a
scientific fact in their lives by induction from
experiment or observation, prate learnedly about
Mr. Darwin's method, which is not inductive
enough, not Baconian enough, forsooth, for them.
But even if practical acquaintance with the process
of scientific investigation is denied them, they may
learn, by the perusal of Mr. Mill's admirable
chapter " On the Deductive Method," that there
are multitudes of scientific inquiries in which the
method of pure induction helps the investigator
but a very little way.

"The mode of investigation," says Mr. Mill, "which, from the proved inapplicability of direct methods of observation and experiment, remains to us as the main source of the knowledge we possess, or can acquire, respecting the conditions and laws of recurrence of the more complex phænomena, is called, in its most general expression, the deductive method, and consists of three operations : the first, one of direct induction ; the second, of ratiocination ; and the third, of verification."

Now, the conditions which have determined the existence of species are not only exceedingly complex, but, so far as the great majority of them are concerned, are necessarily beyond our cognisance. But what Mr. Darwin has attempted to do is in exact accordance with the rule laid down by Mr. Mill; he has endeavoured to determine certain great facts inductively, by observation and experiment ; he has then reasoned from the data thus furnished ; and lastly, he has tested the validity of his ratiocination by comparing his deductions with the observed facts of Nature. Inductively, Mr. Darwin endeavours to prove that species arise in a given way. Deductively, he desires to show that, if they arise in that way, the facts of distribution, development, classification, &c., may be accounted for, *i.e.* may be deduced from their mode of origin, combined with admitted changes in physical geography and climate, during an indefinite period. And this explanation, or coincidence of observed with deduced facts, is, so far as it extends, a verification of the Darwinian view. There is no fault to be found with Mr. Darwin's

method, then ; but it is another question whether
he has fulfilled all the conditions imposed by that
method.   Is it satisfactorily proved, in fact, that
species may be originated by selection ? that there
is such a thing as natural selection ? that none
of the phænomena exhibited by species are inconsistent with the origin of species in this way ?   If
these questions can be answered in the affirmative,
Mr. Darwin's view steps out of the rank of hypotheses into those of proved theories ; but, so long
as the evidence at present adduced falls short of
enforcing that affirmation, so long, to our minds,
must the new doctrine be content to remain among
the former—an extremely valuable, and in the
highest degree probable, doctrine, indeed the only
extant hypothesis which is worth anything in a
scientific point of view ; but still a hypothesis, and
not yet the theory of species.

After much consideration, and with assuredly
no bias against Mr. Darwin's views, it is our clear
conviction that, as the evidence stands, it is not
absolutely proven that a group of animals, having
all the characters exhibited by species in Nature,
has ever been originated by selection, whether
artificial or natural.   Groups having the morphological character of species—distinct and permanent
races in fact—have been so produced over and over
again ; but there is no positive evidence, at present,
that any group of animals has, by variation and
selective breeding, given rise to another group

operation which can be effected by Nature, for man interferes intelligently. Reduced to its elements, this argument implies that an effect produced with trouble by an intelligent agent must, *à fortiori*, be more troublesome, if not impossible, to an unintelligent agent. Even putting aside the question whether Nature, acting as she does according to definite and invariable laws, can be rightly called an unintelligent agent, such a position as this is wholly untenable. Mix salt and sand, and it shall puzzle the wisest of men, with his mere natural appliances, to separate all the grains of sand from all the grains of salt; but a shower of rain will effect the same object in ten minutes. And so, while man may find it tax all his intelligence to separate any variety which arises, and to breed selectively from it, the destructive agencies incessantly at work in Nature, if they find one variety to be more soluble in circumstances than the other, will inevitably, in the long run, eliminate it.

A frequent and a just objection to the Lamarckian hypothesis of the transmutation of species is based upon the absence of transitional forms between many species. But against the Darwinian hypothesis this argument has no force. Indeed, one of the most valuable and suggestive parts of Mr. Darwin's work is that in which he proves, that the frequent absence of transitions is a necessary consequence of his doctrine, and that the stock whence two or more species have sprung, need in no respect be intermediate between these species. If any two species have arisen from a common stock in the same way as the carrier and the pouter, say, have arisen from the rock-pigeon, then the common stock of these two species need be no more intermediate between the two than the rock-pigeon is between the carrier and pouter. Clearly appreciate the force of this analogy, and all the arguments against the origin of species by selection, based on the absence of transitional forms, fall to the ground. And Mr. Darwin's position might, we think, have been even stronger than it is if he had not embarrassed himself with the aphorism, "*Natura non facit saltum*," which turns up so often in his pages. We believe, as we have said above, that Nature does make jumps now and then, and a recognition of the fact is of no small importance in disposing of many minor objections to the doctrine of transmutation.

But we must pause. The discussion of Mr. Darwin's arguments in detail would lead us far beyond the limits within which we proposed, at starting, to confine this article. Our object has been attained if we have given an intelligible, however brief, account of the established facts connected with species, and of the relation of the explanation of those facts offered by Mr. Darwin to the theoretical views held by his predecessors and his contemporaries, and, above all, to the requirements of scientific logic. We have ventured to point out that it does not, as yet, satisfy all those requirements; but we do not hesitate to assert that it is as superior to any preceding or contemporary hypothesis, in the extent of observational and experimental basis on which it rests, in its rigorously scientific method, and in its power of explaining biological phænomena, as was the hypothesis of Copernicus to the speculations of Ptolemy. But the planetary orbits turned out to be not quite circular after all, and, grand as was the service Copernicus rendered to science, Kepler and Newton had to come after him. What if the orbit of Darwinism should be a little too circular? What if species should offer residual phænomena, here and there, not explicable by natural selection? Twenty years hence naturalists may be in a position to say whether this is, or is not, the case; but in either event they will owe the author of "The Origin of Species" an immense debt of gratitude. We should leave a very wrong impression on the reader's mind if we permitted him to suppose that the value of that work depends wholly on the ultimate justification of the theoretical views which it contains. On the contrary, if they were disproved to-morrow, the book would still be the best of its kind—the most compendious statement of well-sifted facts bearing on the doctrine of species that has ever appeared. The chapters on Variation, on the Struggle for

Existence, on Instinct, on Hybridism, on the Imperfection of the Geological Record, on Geographical Distribution, have not only no equals, but, so far as our knowledge goes, no competitors, within the range of biological literature. And viewed as a whole, we do not believe that, since the publication of Von Baer's " Researches on Development," thirty years ago, any work has appeared calculated to exert so large an influence, not only on the future of Biology, but in extending the domination of Science over regions of thought into which she has, as yet, hardly penetrated.

III

CRITICISMS ON "THE ORIGIN OF SPECIES"

[1864]

1. UEBER DIE DARWIN'SCHE SCHÖPFUNGSTHEORIE; EIN VORTRAG, VON A. KÖLLIKER. Leipzig, 1864.
2. EXAMINATION DU LIVRE DE M. DARWIN SUR L'ORIGINE DES ESPÈCES. Par P. FLOURENS. Paris, 1864.

IN the course of the present year several foreign commentaries upon Mr. Darwin's great work have made their appearance. Those who have perused that remarkable chapter of the "Antiquity of Man," in which Sir Charles Lyell draws a parallel between the development of species and that of languages, will be glad to hear that one of the most eminent philologers of Germany, Professor Schleicher, has, independently, published a most instructive and philosophical pamphlet (an excellent notice of which is to be found in the

Reader, for February 27th of this year) supporting similar views with all the weight of his special knowledge and established authority as a linguist. Professor Haeckel, to whom Schleicher addresses himself, previously took occasion, in his splendid monograph on the *Radiolaria*,[1] to express his high appreciation of, and general concordance with, Mr. Darwin's views.

But the most elaborate criticisms of the " Origin of Species" which have appeared are two works of very widely different merit, the one by Professor Kölliker, the well-known anatomist and histologist of Würzburg; the other by M. Flourens, Perpetual Secretary of the French Academy of Sciences.

Professor Kölliker's critical essay " Upon the Darwinian Theory " is, like all that proceeds from the pen of that thoughtful and accomplished writer, worthy of the most careful consideration. It comprises a brief but clear sketch of Darwin's views, followed by an enumeration of the leading difficulties in the way of their acceptance; difficulties which would appear to be insurmountable to Professor Kölliker, inasmuch as he proposes to replace Mr. Darwin's Theory by one which he terms the " Theory of Heterogeneous Generation." We shall proceed to consider first the destructive, and secondly, the constructive portion of the essay.

[1] *Die Radiolarien : eine Monographie*, p. 231.

We regret to find ourselves compelled to dissent very widely from many of Professor Kölliker's remarks; and from none more thoroughly than from those in which he seeks to define what we may term the philosophical position of Darwinism.

"Darwin," says Professor Kölliker, "is, in the fullest sense of the word, a Teleologist. He says quite distinctly (First Edition, pp. 199, 200) that every particular in the structure of an animal has been created for its benefit, and he regards the whole series of animal forms only from this point of view."

And again :

"7. The teleological general conception adopted by Darwin is a mistaken one.

"Varieties arise irrespectively of the notion of purpose, or of utility, according to general laws of Nature, and may be either useful, or hurtful, or indifferent.

"The assumption that an organism exists only on account of some definite end in view, and represents something more than the incorporation of a general idea, or law, implies a one-sided conception of the universe. Assuredly, every organ has, and every organism fulfils, its end, but its purpose is not the condition of its existence. Every organism is also sufficiently perfect for the purpose it serves, and in that, at least, it is useless to seek for a cause of its improvement."

It is singular how differently one and the same book will impress different minds. That which struck the present writer most forcibly on his first perusal of the "Origin of Species" was the conviction that Teleology, as commonly understood, had received its deathblow at Mr. Darwin's hands. For the teleological argument runs thus : an organ or organism (A) is precisely fitted to perform a function or purpose (B); therefore it was specially constructed to perform that function. In Paley's famous illustration, the adaptation of all the parts of the watch to the function, or purpose, of showing the time, is held to be evidence that the watch was specially contrived to that end; on the ground, that the only cause we know of, competent to produce such an effect as a watch which shall keep time, is a contriving intelligence adapting the means directly to that end.

Suppose, however, that any one had been able to show that the watch had not been made directly by any person, but that it was the result of the modification of another watch which kept time but poorly; and that this again had proceeded from a structure which could hardly be called a watch at all—seeing that it had no figures on the dial and the hands were rudimentary; and that going back and back in time we came at last to a re-volving barrel as the earliest traceable rudiment of the whole fabric. And imagine that it had been possible to show that all these changes had resulted, first, from a tendency of the structure to vary indefinitely; and secondly, from something in the surrounding world which helped all variations in the direction of an accurate time-keeper, and checked all those in other directions; then it is obvious that the force of Paley's argument would be gone. For it would be demonstrated that an apparatus thoroughly well adapted to a particular purpose might be the result of a method of trial and error worked by unintelligent agents, as well as of the direct application of the means appropriate to that end, by an intelligent agent.

Now it appears to us that what we have here, for illustration's sake, supposed to be done with the watch, is exactly what the establishment of Darwin's Theory will do for the organic world. For the notion that every organism has been created as it is and launched straight at a purpose, Mr. Darwin substitutes the conception of something which may fairly be termed a method of trial and error. Organisms vary incessantly; of these variations the few meet with surrounding conditions which suit them and thrive ; the many are unsuited and become extinguished.

According to Teleology, each organism is like a rifle bullet fired straight at a mark; according to Darwin, organisms are like grapeshot of which one hits something and the rest fall wide.

For the teleologist an organism exists because it was made for the conditions in which it is found ; for the Darwinian an organism exists because, out of many of its kind, it is the only one which has been able to persist in the conditions in which it is found.

Teleology implies that the organs of every organism are perfect and cannot be improved ; the Darwinian theory simply affirms that they work

well enough to enable the organism to hold its
own against such competitors as it has met with,
but admits the possibility of indefinite improvement. But an example may bring into clearer
light the profound opposition between the ordinary
teleological, and the Darwinian, conception.

Cats catch mice, small birds and the like, very
well. Teleology tells us that they do so because
they were expressly constructed for so doing—that
they are perfect mousing apparatuses, so perfect
and so delicately adjusted that no one of their organs could be altered, without the change involving
the alteration of all the rest. Darwinism affirms
on the contrary, that there was no express construction concerned in the matter; but that among
the multitudinous variations of the Feline stock,
many of which died out from want of power to
resist opposing influences, some, the cats, were
better fitted to catch mice than others, whence
they throve and persisted, in proportion to the
advantage over their fellows thus offered to them.

Far from imagining that cats exist *in order* to
catch mice well, Darwinism supposes that cats exist
because they catch mice well—mousing being not
the end, but the condition, of their existence. And
if the cat type has long persisted as we know it,
the interpretation of the fact upon Darwinian
principles would be, not that the cats have remained invariable, but that such varieties as have
incessantly occurred have been, on the whole, less

fitted to get on in the world than the existing stock.

If we apprehend the spirit of the " Origin of Species " rightly, then, nothing can be more entirely and absolutely opposed to Teleology, as it is commonly understood, than the Darwinian Theory. So far from being a " Teleologist in the fullest sense of the word," we should deny that he is a Teleologist in the ordinary sense at all ; and we should say that, apart from his merits as a naturalist, he has rendered a most remarkable service to philosophical thought by enabling the student of Nature to recognise, to their fullest extent, those adaptations to purpose which are so striking in the organic world, and which Teleology has done good service in keeping before our minds, without being false to the fundamental principles of a scientific conception of the universe. The apparently diverging teachings of the Teleologist and of the Morphologist are reconciled by the Darwinian hypothesis.

But leaving our own impressions of the " Origin of Species," and turning to those passages especially cited by Professor Kölliker, we cannot admit that they bear the interpretation he puts upon them. Darwin, if we read him rightly, does *not* affirm that every detail in the structure of an animal has been created for its benefit.   His words are (p. 199) :—

" The foregoing remarks lead me to say a few words on the protest lately made by some naturalists against the utilitarian doctrine that every detail of structure has been produced for the

good of its possessor. They believe that very many structures
have been created for beauty in the eyes of man, or for mere
variety. This doctrine, if true, would be absolutely fatal to my
theory—yet I fully admit that many structures are of no direct
use to their possessor."

And after sundry illustrations and qualifications,
he concludes (p. 200) :—

"Hence every detail of structure in every living creature
(making some little allowance for the direct action of physical
conditions) may be viewed either as having been of special use
to some ancestral form, or as being now of special use to the
descendants of this form—either directly, or indirectly, through
the complex laws of growth."

But it is one thing to say, Darwinically, that
every detail observed in an animal's structure is
of use to it, or has been of use to its ancestors ;
and quite another to affirm, teleologically, that
every detail of an animal's structure has been
created for its benefit. On the former hypothesis,
for example, the teeth of the fœtal *Balœna* have a
meaning ; on the latter, none. So far as we are
aware, there is not a phrase in the " Origin of
Species " inconsistent with Professor Kölliker's
position, that " varieties arise irrespectively of the
notion of purpose, or of utility, according to general
laws of Nature, and may be either useful, or hurtful, or indifferent."

On the contrary, Mr. Darwin writes (Summary
of Chap. V.) :—

"Our ignorance of the laws of variation is profound. Not in
one case out of a hundred can we pretend to assign any reason
why this or that part varies more or less from the same part in

the parents. . . The external conditions of life, as climate and food, &c., seem to have induced some slight modifications. Habit, in producing constitutional differences, and use, in strengthening, and disuse, in weakening and diminishing organs, seem to have been more potent in their effects."

And finally, as if to prevent all possible misconception, Mr. Darwin concludes his Chapter on Variation with these pregnant words :—

" Whatever the cause may be of each slight difference in the offspring from their parents—and a cause for each must exist—it is the steady accumulation, through natural selection of such differences, when beneficial to the individual, that gives rise to all the more important modifications of structure, by which the innumerable beings on the face of the earth are enabled to struggle with each other, and the best adapted to survive."

We have dwelt at length upon this subject, because of its great general importance, and because we believe that Professor Kölliker's criticisms on this head are based upon a misapprehension of Mr. Darwin's views—substantially they appear to us to coincide with his own. The other objections which Professor Kölliker enumerates and discusses are the following : [1]—

"1. No transitional forms between existing species are known ; and known varieties, whether selected or spontaneous, never go so far as to establish new species."

To this Professor Kölliker appears to attach some weight. He makes the suggestion that the

---

[1] Space will not allow us to give Professor Kölliker's arguments in detail ; our readers will find a full and accurate version of them in the *Reader* for August 13th and 20th, 1864.

short-faced tumbler pigeon may be a pathological product.

"2. No transitional forms of animals are met with among the organic remains of earlier epochs."

Upon this, Professor Kölliker remarks that the absence of transitional forms in the fossil world, though not necessarily fatal to Darwin's views, weakens his case.

"3. The struggle for existence does not take place."

To this objection, urged by Pelzeln, Kölliker, very justly, attaches no weight.

"4. A tendency of organisms to give rise to useful varieties, and a natural selection, do not exist.

"The varieties which are found arise in consequence of manifold external influences, and it is not obvious why they all, or partially, should be particularly useful. Each animal suffices for its own ends, is perfect of its kind, and needs no further development. Should, however, a variety be useful and even maintain itself, there is no obvious reason why it should change any further. The whole conception of the imperfection of organisms and the necessity of their becoming perfected is plainly the weakest side of Darwin's Theory, and a *pis aller* (Nothbehelf) because Darwin could think of no other principle by which to explain the metamorphoses which, as I also believe, have occurred."

Here again we must venture to dissent completely from Professor Kölliker's conception of Mr. Darwin's hypothesis. It appears to us to be one of the many peculiar merits of that hypothesis that it involves no belief in a necessary and continual progress of organisms.

Again, Mr. Darwin, if we read him aright,

assumes no special tendency of organisms to give rise to useful varieties, and knows nothing of needs of development, or necessity of perfection. What he says is, in substance: All organisms vary. It is in the highest degree improbable that any given variety should have exactly the same relations to surrounding conditions as the parent stock. In that case it is either better fitted (when the variation may be called useful), or worse fitted, to cope with them. If better, it will tend to supplant the parent stock; if worse, it will tend to be extinguished by the parent stock.

If (as is hardly conceivable) the new variety is so perfectly adapted to the conditions that no improvement upon it is possible,—it will persist, because, though it does not cease to vary, the varieties will be inferior to itself.

If, as is more probable, the new variety is by no means perfectly adapted to its conditions, but only fairly well adapted to them, it will persist, so long as none of the varieties which it throws off are better adapted than itself.

On the other hand, as soon as it varies in a useful way, i.e. when the variation is such as to adapt it more perfectly to its conditions, the fresh variety will tend to supplant the former.

So far from a gradual progress towards perfection forming any necessary part of the Darwinian creed, it appears to us that it is perfectly consistent with indefinite persistence in one state, or with a gradual retrogression. Suppose, for example, a return of the glacial epoch and a spread of polar climatal conditions over the whole globe. The operation of natural selection under these circumstances would tend, on the whole, to the weeding out of the higher organisms and the cherishing of the lower forms of life. Cryptogamic vegetation would have the advantage over Phanerogamic; *Hydrozoa* over Corals; *Crustacea* over *Insecta*, and *Amphipoda* and *Isopoda* over the higher *Crustacea;* Cetaceans and Seals over the *Primates;* the civilisation of the Esquimaux over that of the European.

"5. Pelzeln has also objected that if the later organisms have proceeded from the earlier, the whole developmental series, from the simplest to the highest, could not now exist; in such a case the simpler organisms must have disappeared."

To this Professor Kölliker replies, with perfect justice, that the conclusion drawn by Pelzeln does not really follow from Darwin's premises, and that, if we take the facts of Palæontology as they stand, they rather support than oppose Darwin's theory.

"6. Great weight must be attached to the objection brought forward by Huxley, otherwise a warm supporter of Darwin's hypothesis, that we know of no varieties which are sterile with one another, as is the rule among sharply distinguished animal forms.

"If Darwin is right, it must be demonstrated that forms may be produced by selection, which, like the present sharply distinguished animal forms, are infertile, when coupled with one another, and this has not been done."

The weight of this objection is obvious ; but our ignorance of the conditions of fertility and sterility, the want of carefully conducted experiments extending over long series of years, and the strange anomalies presented by the results of the cross-fertilisation of many plants, should all, as Mr. Darwin has urged, be taken into account in considering it.

The seventh objection is that we have already discussed (*supra* p. 82).

The eighth and last stands as follows :—

" 8. The developmental theory of Darwin is not needed to enable us to understand the regular harmonious progress of the complete series of organic forms from the simpler to the more perfect.

" The existence of general laws of Nature explains this harmony, even if we assume that all beings have arisen separately and independent of one another. Darwin forgets that inorganic nature, in which there can be no thought of genetic connexion of forms, exhibits the same regular plan, the same harmony, as the organic world ; and that, to cite only one example, there is as much a natural system of minerals as of plants and animals."

We do not feel quite sure that we seize Professor Kölliker's meaning here, but he appears to suggest that the observation of the general order and harmony which pervade inorganic nature, would lead us to anticipate a similar order and harmony in the organic world. And this is no doubt true, but it by no means follows that the particular order and harmony observed among them should be that which we see. Surely the stripes of dun horses, and the teeth of the fœtal *Balœna*, are not explained by the "existence of general laws of Nature." Mr. Darwin endeavours to explain the exact order of organic nature which exists; not the mere fact that there is some order.

And with regard to the existence of a natural system of minerals; the obvious reply is that there may be a natural classification of any objects—of stones on a sea-beach, or of works of art; a natural classification being simply an assemblage of objects in groups, so as to express their most important and fundamental resemblances and differences. No doubt Mr. Darwin believes that those resemblances and differences upon which our natural systems or classifications of animals and plants are based, are resemblances and differences which have been produced genetically, but we can discover no reason for supposing that he denies the existence of natural classifications of other kinds.

And, after all, is it quite so certain that a genetic relation may not underlie the classification of minerals? The inorganic world has not always been what we see it. It has certainly had its metamorphoses, and, very probably, a long "Entwickelungsgeschichte" out of a nebular blastema. Who knows how far that amount of likeness among sets of minerals, in virtue of which they are now grouped into families and orders, may not be the expression of the common conditions to which that particular patch of nebulous fog, which may have been constituted by their atoms, and of which they may be, in the strictest sense, the descendants, was subjected ?

It will be obvious from what has preceded, that we do not agree with Professor Kölliker in thinking the objections which he brings forward so weighty as to be fatal to Darwin's view. But even if the case were otherwise, we should be unable to accept the " Theory of Heterogeneous Generation " which is offered as a substitute. That theory is thus stated :—

"The fundamental conception of this hypothesis is, that, under the influence of a general law of development, the germs of organisms produce others different from themselves. This might happen (1) by the fecundated ova passing, in the course of their development, under particular circumstances, into higher forms ; (2) by the primitive and later organisms producing other organisms without fecundation, out of germs or eggs (Parthenogenesis)."

In favour of this hypothesis, Professor Kölliker adduces the well-known facts of Agamogenesis, or "alternate generation " ; the extreme dissimilarity of the males and females of many animals ; and of the males, females, and neuters .of those insects which live in colonies : and he defines its relations to the Darwinian theory as follows :—

" It is obvious that my hypothesis is apparently very similar to Darwin's, inasmuch as I also consider that the various forms of animals have proceeded directly from one another. My hypothesis of the creation of organisms by heterogeneous generation, however, is distinguished very essentially from Darwin's by the entire absence of the principle of useful variations and their natural selection : and my fundamental conception is this, that a great plan of development lies at the foundation of the origin of the whole organic world, impelling the simpler forms to more and more complex developments. How this law operates, what influences determine the development of the eggs and germs, and impel them to assume constantly new forms, I naturally cannot pretend to say ; but I can at least adduce the great analogy of the alternation of generations. If a *Bipinnaria,* a *Brachiolaria,* a *Pluteus,* is competent to produce the Echinoderm, which is so widely different from it ; if a hydroid polype can produce the higher Medusa ; if the vermiform Trematode 'nurse' can develop within itself the very unlike *Cercaria,* it will not appear impossible that the egg, or ciliated embryo, of a sponge, for once, under special conditions, might become a hydroid polype, or the embryo of a Medusa, an Echinoderm."

It is obvious, from these extracts, that Professor Kölliker's hypothesis is based upon the supposed existence of a close analogy between the phænomena of Agamogenesis and the production of new species from pre-existing ones. But is the analogy a real one ? We think that it is not, and, by the hypothesis cannot be

For what are the phænomena of Agamogenesis, stated generally ? An impregnated egg develops into a sexless form, A ; this gives rise, non-sexually, to a second form or forms, B, more or less different from A. B may multiply non-sexually again ; in the simpler cases, however, it does not, but, acquiring sexual characters, produces impregnated eggs from whence A, once more, arises.

No case of Agamogenesis is known in which *when A differs widely from B*, it is itself capable of sexual propagation.   No case whatever is known in which the progeny of B, by sexual generation, is other than a reproduction of A.

But if this be a true statement of the nature of the process of Agamogenesis, how can it enable us to comprehend the production of new species from already existing ones?   Let us suppose Hyænas to have preceded Dogs, and to have produced the latter in this way.   Then the Hyæna will represent A, and the Dog, B.   The first difficulty that presents itself is that the Hyæna must be non-sexual, or the process will be wholly without analogy in the world of Agamogenesis.   But passing over this difficulty, and supposing a male and female Dog to be produced at the same time from the Hyæna stock, the progeny of the pair, if the analogy of the simpler kinds of Agamogenesis[1] is to be followed, should be a litter, not of puppies, but of young Hyænas.   For the Agamogenetic series is

---

[1] If, on the contrary, we follow the analogy of the more complex forms of Agamogenesis, such as that exhibited by some *Trematoda* and by the *Aphides*, the Hyæna must produce, non-sexually, a brood of sexless Dogs, from which other sexless Dogs must proceed.   At the end of a certain number of terms of the series, the Dogs would acquire sexes and generate young; but these young would be, not Dogs, but Hyænas   In fact, we have demonstrated, in Agamogenetic phænomena, that inevitable recurrence to the original type, which is asserted to be true of variations in general, by Mr. Darwin's opponents; and which, if the assertion could be changed into a demonstration, would, in fact, be fatal to his hypothesis.

always, as we have seen, A : B : A : B, &c. ; whereas, for the production of a new species, the series must be A : B : B : B, &c. The production of new species, or genera, is the extreme permanent divergence from the primitive stock. All known Agamogenetic processes, on the other hand, end in a complete return to the primitive stock. How then is the production of new species to be rendered intelligible by the analogy of Agamogenesis ?

The other alternative put by Professor Kölliker —the passage of fecundated ova in the course of their development into higher forms—would, if it occurred, be merely an extreme case of variation in the Darwinian sense, greater in degree than, but perfectly similar in kind to, that which occurred when the well-known Ancon Ram was developed from an ordinary Ewe's ovum. Indeed we have always thought that Mr. Darwin has unnecessarily hampered himself by adhering so strictly to his favourite " Natura non facit saltum." We greatly suspect that she does make considerable jumps in the way of variation now and then, and that these saltations give rise to some of the gaps which appear to exist in the series of known forms.

Strongly and freely as we have ventured to disagree with Professor Kölliker, we have always done so with regret, and we trust without violating that respect which is due, not only to his scientific eminence and to the careful study which he has

devoted to the subject, but to the perfect fairness
of his argumentation, and the generous appreciation
of the worth of Mr. Darwin's labours which he
always displays.   It would be satisfactory to be
able to say as much for M. Flourens.

But the Perpetual Secretary of the French
Academy of Sciences deals with Mr. Darwin as the
first Napoleon would have treated an "idéologue;"
and while displaying a painful weakness of logic
and shallowness of information, assumes a tone of
authority, which always touches upon the ludicrous,
and sometimes passes the limits of good breeding.
For example (p. 56) :—

"M. Darwin continue : 'Aucune distinction absolue n'a été
et ne peut être établie entre les espèces et les variétés.' Je vous
ai déjà dit que vous vous trompiez ; une distinction absolue
sépare les variétés d'avec les espèces."

" *Je vous ai déjà dit ;* moi, M. le Secrétaire per-
pétuel de l'Académie des Sciences : et vous

   "'Qui n'êtes rien,
      Pas même Académicien ;'

what do you mean by asserting the contrary ? "
Being devoid of the blessings of an Academy in
England, we are unaccustomed to see our ablest
men treated in this fashion, even by a "Perpetual
Secretary."

Or again, considering that if there is any one
quality of Mr. Darwin's work to which friends and
foes have alike borne witness, it is his candour and

fairness in admitting and discussing objections, what is to be thought of M. Flourens' assertion, that

"M. Darwin ne cite que les auteurs qui partagent ses opinions." (P. 40.)

Once more (p. 65) :—

"Enfin l'ouvrage de M. Darwin a paru. On ne peut qu'être frappé du talent de l'auteur. Mais que d'idées obscures, que d'idées fausses! Quel jargon métaphysique jeté mal à propos dans l'histoire naturelle, qui tombe dans le galimatias dès qu'elle sort des idées claires, des idées justes! Quel langage prétentieux et vide! Quelles personnifications puériles et surannées! O lucidité! O solidité de l'esprit Français, que devenez-vous?"

"Obscure ideas," "metaphysical jargon," "pretentious and empty language," "puerile and superannuated personifications." Mr. Darwin has many and hot opponents on this side of the Channel and in Germany, but we do not recollect to have found precisely these sins in the long catalogue of those hitherto laid to his charge. It is worth while, therefore, to examine into these discoveries effected solely by the aid of the "lucidity and solidity" of the mind of M. Flourens.

According to M. Flourens, Mr. Darwin's great error is that he has personified Nature (p. 10), and further that he has

"imagined a natural selection : he imagines afterwards that this power of selecting (*pouvoir d'élire*) which he gives to Nature is similar to the power of man. These two suppositions ad-

H 2

mitted, nothing stops him : he plays with Nature as he likes, and makes her do all he pleases." (P. 6.)

And this is the way M. Flourens extinguishes natural selection :

"Voyons donc encore une fois, ce qu'il peut y avoir de fondé dans ce qu'on nomme *élection naturelle.*

" *L'élection naturelle* n'est sous un autre nom que la nature. Pour un être organisé, la nature n'est que l'organisation, ni plus ni moins.

"Il faudra donc aussi personnifier *l'organisation*, et dire que *l'organisation* choisit *l'organisation*. *L'élection naturelle* est cette *forme substantielle* dont on jouait autrefois avec tant de facilité. Aristote disait que 'Si l'art de bâtir était dans le bois, cet art agirait comme la nature.' A la place de *l'art de bâtir* M. Darwin met *l'élection naturelle*, et c'est tout un : l'un n'est pas plus chimérique que l'autre." (P. 31.)

And this is really all that M. Flourens can make of Natural Selection. We have given the original, in fear lest a translation should be regarded as a travesty ; but with the original before the reader, we may try to analyse the passage. "For an organised being, Nature is only organisation, neither more nor less."

Organised beings then have absolutely no relation to inorganic nature : a plant does not depend on soil or sunshine, climate, depth in the ocean, height above it; the quantity of saline matters in water have no influence upon animal life ; the substitution of carbonic acid for oxygen in our atmosphere would hurt nobody ! That these are absurdities no one should know better than M. Flourens; but they are logical deductions from the assertion just quoted, and from the further statement that natural selection means only that " organisation chooses and selects organisation."

For if it be once admitted (what no sane man denies) that the chances of life of any given organism are increased by certain conditions (A) and diminished by their opposites (B), then it is mathematically certain that any change of conditions in the direction of (A) will exercise a selective influence in favour of that organism, tending to its increase and multiplication, while any change in the direction of (B) will exercise a selective influence against that organism, tending to its decrease and extinction.

Or, on the other hand, conditions remaining the same, let a given organism vary (and no one doubts that they do vary) in two directions : into one form (*a*) better fitted to cope with these conditions than the original stock, and a second (*b*) less well adapted to them. Then it is no less certain that the conditions in question must exercise a selective influence in favour of (*a*) and against (*b*), so that (*a*) will tend to predominance, and (*b*) to extirpation.

That M. Flourens should be unable to perceive the logical necessity of these simple arguments, which lie at the foundation of all Mr. Darwin's reasoning ; that he should confound an irrefragable

deduction from the observed relations of organisms
to the conditions which lie around them, with a
metaphysical "forme substantielle," or a chimerical
personification of the powers of Nature, would be
incredible, were it not that other passages of his
work leave no room for doubt upon the subject.

"On imagine une *élection naturelle* que, pour plus de ménage-
ment, on me dit être *inconsciente*, sans s'apercevoir que le contre-
sens littéral est précisément là : *élection inconsciente.*" (P. 52.)
"J'ai déjà dit ce qu'il faut penser de *l'élection naturelle.* Ou
l'élection naturelle n'est rien, ou c'est la nature : mais la nature
douée *d'élection*, mais la nature personnifiée : dernière erreur du
dernier siècle : Le xixe ne fait plus de personnifications." (P.
53.)

M. Flourens cannot imagine an unconscious
selection—it is for him a contradiction in terms.
Did M. Flourens ever visit one of the prettiest
watering-places of "la belle France," the Baie
d'Arcachon ? If so, he will probably have passed
through the district of the Landes, and will have
had an opportunity of observing the formation of
"dunes" on a grand scale. What are these
"dunes"? The winds and waves of the Bay of
Biscay have not much consciousness, and yet they
have with great care "selected," from among an
infinity of masses of silex of all shapes and sizes,
which have been submitted to their action, all the
grains of sand below a certain size, and have
heaped them by themselves over a great area.
This sand has been "unconsciously selected" from

amidst the gravel in which it first lay with as much precision as if man had "consciously selected" it by the aid of a sieve. Physical Geology is full of such selections—of the picking out of the soft from the hard, of the soluble from the insoluble, of the fusible from the infusible, by natural agencies to which we are certainly not in the habit of ascribing consciousness.

But that which wind and sea are to a sandy beach, the sum of influences, which we term the "conditions of existence," is to living organisms. The weak are sifted out from the strong. A frosty night "selects" the hardy plants in a plantation from among the tender ones as effectually as if it were the wind, and they, the sand and pebbles, of our illustration; or, on the other hand, as if the intelligence of a gardener had been operative in cutting the weaker organisms down. The thistle, which has spread over the Pampas, to the destruction of native plants, has been more effectually "selected" by the unconscious operation of natural conditions than if a thousand agriculturists had spent their time in sowing it.

It is one of Mr. Darwin's many great services to Biological science that he has demonstrated the significance of these facts. He has shown that—given variation and given change of conditions—the inevitable result is the exercise of such an influence upon organisms that one is helped and another is impeded; one tends to predominate, another to disappear; and thus the living world bears within itself, and is surrounded by, impulses towards incessant change.

But the truths just stated are as certain as any other physical laws, quite independently of the truth, or falsehood, of the hypothesis which Mr. Darwin has based upon them; and that M. Flourens, missing the substance and grasping at a shadow, should be blind to the admirable exposition of them, which Mr. Darwin has given, and see nothing there but a "dernière erreur du dernier siècle"—a personification of Nature—leads us indeed to cry with him: "O lucidité! O solidité de l'esprit Francais, que devenez-vous?"

M. Flourens has, in fact, utterly failed to comprehend the first principles of the doctrine which he assails so rudely. His objections to details are of the old sort, so battered and hackneyed on this side of the Channel, that not even a Quarterly Reviewer could be induced to pick them up for the purpose of pelting Mr. Darwin over again. We have Cuvier and the mummies; M. Roulin and the domesticated animals of America; the difficulties presented by hybridism and by Palæontology; Darwinism a *rifacciamento* of De Maillet and Lamarck; Darwinism a system without a commencement, and its author bound to believe in M. Pouchet, &c. &c. How one knows it all by heart, and with what relief one reads at p. 65—

"Je laisse M. Darwin!"

But we cannot leave M. Flourens without calling our readers' attention to his wonderful tenth chapter, " De la Préexistence des Germes et de l'Epigénèse," which opens thus :—

"Spontaneous generation is only a chimæra. This point established, two hypotheses remain : that of *pre-existence* and that of *epigenesis*. The one of these hypotheses has as little foundation as the other." (P. 163.)

"The doctrine of *epigenesis* is derived from Harvey : following by ocular inspection the development of the new being in the Windsor does, he saw each part appear successively, and taking the moment of *appearance* for the moment of *formation* he imagined *epigenesis*." (P. 165.)

On the contrary, says M. Flourens (p. 167),

"The new being is formed at a stroke (*tout d'un coup*), as a whole, instantaneously ; it is not formed part by part, and at different times. It is formed at once at the single *individual* moment at which the conjunction of the male and female elements takes place."

It will be observed that M. Flourens uses language which cannot be mistaken. For him, the labours of Von Baer, of Rathke, of Coste, and their contemporaries and successors in Germany, France, and England, are non-existent : and, as Darwin " *imagina* " natural selection, so Harvey " *imagina* " that doctrine which gives him an even greater claim to the veneration of posterity than his better known discovery of the circulation of the blood.

Language such as that we have quoted is, in fact, so preposterous, so utterly incompatible with anything but absolute ignorance of some of the best established facts, that we should have passed it over in silence had it not appeared to afford some clue to M. Flourens' unhesitating, *à priori*, repudiation of all forms of the doctrine of progressive modification of living beings. He whose mind remains uninfluenced by an acquaintance with the phænomena of development, must indeed lack one of the chief motives towards the endeavour to trace a genetic relation between the different existing forms of life. Those who are ignorant of Geology, find no difficulty in believing that the world was made as it is ; and the shepherd, untutored in history, sees no reason to regard the green mounds which indicate the site of a Roman camp, as aught but part and parcel of the primæval hill side. So M. Flourens, who believes that embryos are formed " tout d'un coup," naturally finds no difficulty in conceiving that species came into existence in the same way.

IV

THE GENEALOGY OF ANIMALS [1]

[1869]

CONSIDERING that Germany now takes the lead of
the world in scientific investigation, and particularly in biology, Mr. Darwin must be well pleased
at the rapid spread of his views among some of
the ablest and most laborious of German
naturalists.

Among these, Professor Haeckel, of Jena, is the
Coryphæus. I know of no more solid and important contributions to biology in the past seven
years than Haeckel's work on the "Radiolaria,"
and the researches of his distinguished colleague
Gegenbaur, in vertebrate anatomy; while in
Haeckel's "Generelle Morphologie" there is all
the force, suggestiveness, and, what I may term

[1] *The Natural History of Creation.* By Dr. Ernst Haeckel.
[*Natürliche Schöpfungs-Geschichte.*—Von Dr. Ernst Haeckel,
Professor an der Universität Jena.] Berlin, 1868.

the systematising power, of Oken, without his extravagance. The "Generelle Morphologie" is, in fact, an attempt to put the Doctrine of Evolution, so far as it applies to the living world, into a logical form ; and to work out its practical applications to their final results. The work before us, again, may be said to be an exposition of the "Generelle Morphologie" for an educated public, consisting, as it does, of the substance of a series of lectures delivered before a mixed audience at Jena, in the session 1867–8.

"The Natural History of Creation,"—or, as Professor Haeckel admits it would have been better to call his work, "The History of the Development or Evolution of Nature,"—deals, in the first six lectures, with the general and historical aspects of the question and contains a very interesting and lucid account of the views of Linnæus, Cuvier, Agassiz, Goethe, Oken, Kant, Lamarck, Lyell, and Darwin, and of the historical filiation of these philosophers.

The next six lectures are occupied by a well-digested statement of Mr. Darwin's views. The thirteenth lecture discusses two topics which are not touched by Mr. Darwin, namely, the origin of the present form of the solar system, and that of living matter. Full justice is done to Kant, as the originator of that "cosmic gas theory," as the Germans somewhat quaintly call it, which is commonly ascribed to Laplace. With respect to spontaneous generation, while admitting that there is no experimental evidence in its favour, Professor Haeckel denies the possibility of disproving it, and points out that the assumption that it has occurred is a necessary part of the doctrine of Evolution. The fourteenth lecture, on " Schöpfungs-Perioden und Schöpfungs-Urkunden," answers pretty much to the famous disquisition on the "Imperfection of the Geological Record" in the "Origin of Species."

The following five lectures contain the most original matter of any, being devoted to "Phylogeny," or the working out of the details of the process of Evolution in the animal and vegetable kingdoms, so as to prove the line of descent of each group of living beings, and to furnish it with its proper genealogical tree, or "phylum."

The last lecture considers objections and sums up the evidence in favour of biological Evolution.

I shall best testify to my sense of the value of the work thus briefly analysed if I now proceed to note down some of the more important criticisms which have been suggested to me by its perusal.

I. In more than one place, Professor Haeckel enlarges upon the service which the "Origin of Species" has done, in favouring what he terms the "causal or mechanical" view of living nature as opposed to the "teleological or vitalistic" view. And no doubt it is quite true that the doctrine of Evolution is the most formidable opponent of all the commoner and coarser forms of Teleology. But perhaps the most remarkable service to the philosophy of Biology rendered by Mr. Darwin is the reconciliation of Teleology and Morphology, and the explanation of the facts of both which his views offer.

The Teleology which supposes that the eye, such as we see it in man or one of the higher *Verte-brata*, was made with the precise structure which it exhibits, for the purpose of enabling the animal which possesses it to see, has undoubtedly received its death-blow. Nevertheless it is necessary to remember that there is a wider Teleology, which is not touched by the doctrine of Evolution, but is actually based upon the fundamental proposition of Evolution. That proposition is, that the whole world, living and not living, in the result of the mutual interaction, according to definite laws, of the forces possessed by the molecules of which the primitive nebulosity of the universe was composed. If this be true, it is no less certain that the existing world lay, potentially, in the cosmic vapour ; and that a sufficient intelligence could, from a knowledge of the properties of the molecules of that vapour, have predicted, say the state of the Fauna of Britain in 1869, with as much certainty as one can say what will happen to the vapour of the breath in a cold winter's day.

Consider a kitchen clock, which ticks loudly, shows the hours, minutes, and seconds, strikes, cries "cuckoo!" and perhaps shows the phases of the moon. When the clock is wound up, all the phenomena which it exhibits are potentially contained in its mechanism, and a clever clockmaker could predict all it will do after an examination of its structure.

If the evolution theory is correct, the molecular structure of the cosmic gas stands in the same relation to the phenomena of the world as the structure of the clock to its phenomena.

Now let us suppose a death-watch, living in the clock-case, to be a learned and intelligent student of its works. He might say, " I find here nothing but matter and force and pure mechanism from beginning to end," and he would be quite right. But if he drew the conclusion that the clock was not contrived for a purpose, he would be quite wrong. On the other hand, imagine another death-watch of a different turn of mind. He, listening to the monotonous " tick ! tick !" so exactly like his own, might arrive at the conclusion that the clock was itself a monstrous sort of death-watch, and that its final cause and purpose was to tick. How easy to point to the clear relation of the whole mechanism to the pendulum, to the fact that the one thing the clock did always and without intermission was to tick, and that all the rest of its phenomena were intermittent and subordinate to ticking ! For all this, it is certain that kitchen clocks are not contrived for the purpose of making a ticking noise.

Thus the teleological theorist would be as wrong as the mechanical theorist, among our death-watches; and, probably, the only death-watch who would be right would be the one who should maintain that the sole thing death-watches could be sure about was the nature of the clock-works and the way they move; and that the purpose of the clock lay wholly beyond the purview of beetle faculties.

Substitute "cosmic vapour" for "clock," and "molecules" for "works," and the application of the argument is obvious. The teleological and the mechanical views of nature are not, necessarily, mutually exclusive. On the contrary, the more purely a mechanist the speculator is, the more firmly does he assume a primordial molecular arrangement, of which all the phenomena of the universe are the consequences; and the more completely is he thereby at the mercy of the teleologist, who can always defy him to disprove that this primordial molecular arrangement was not intended to evolve the phenomena of the universe. On the other hand, if the teleologist assert that this, that, or the other result of the working of any part of the mechanism of the universe is its purpose and final cause, the mechanist can always inquire how he knows that it is more than an unessential incident

—the mere ticking of the clock, which he mistakes for its function. And there seems to be no reply to this inquiry, any more than to the further, not irrational, question, why trouble one's self about matters which are out of reach, when the working of the mechanism itself, which is of infinite practical importance, affords scope for all our energies?

Professor Haeckel has invented a new and convenient name "Dysteleology," for the study of the "purposelessnesses" which are observable in living organisms—such as the multitudinous cases of rudimentary and apparently useless structures. I confess, however, that it has often appeared to me that the facts of Dysteleology cut two ways. If we are to assume, as evolutionists in general do, that useless organs atrophy, such cases as the existence of lateral rudiments of toes, in the foot of a horse, place us in a dilemma. For, either these rudiments are of no use to the animal, in which case, considering that the horse has existed in its present form since the Pliocene epoch, they surely ought to have disappeared; or they are of some use to the animal, in which case they are of no use as arguments against Teleology. A similar, but still stronger, argument may be based upon the existence of teats, and even functional mammary glands, in male mammals. Numerous cases of "Gynæcomasty," or functionally active breasts in men, are on record, though there is no mammalian species whatever in which the male normally suckles the young. Thus, there can be little doubt that the mammary gland was as apparently useless in the remotest male mammalian ancestor of man as in living men, and yet it has not disappeared. Is it then still profitable to the male organism to retain it? Possibly; but in that case its dysteleological value is gone.[1]

II. Professor Haeckel looks upon the causes which have led to the present diversity of living nature as twofold. Living matter, he tells us, is urged by two impulses: a centripetal, which tends to preserve and transmit the specific form, and which he identifies with heredity; and a centrifugal, which results from the tendency of external conditions to modify the organism and effect its adaptation to themselves. The internal impulse is conservative, and tends to the preservation of specific, or individual, form; the external impulse is metamorphic, and tends to the modification of specific, or individual, form.

In developing his views upon this subject, Professor Haeckel introduces qualifications which disarm some of the criticisms I should have been disposed to offer; but I think that his method of stating the case has the inconvenience of tending to leave out of sight the important fact—which is a cardinal point in the Darwinian hypothesis—

[1] [The recent discovery of the important part played by the Thyroid gland should be a warning to all speculators about useless organs. 1893.]

that the tendency to vary, in a given organism, may have nothing to do with the external conditions to which that individual organism is exposed, but may depend wholly upon internal conditions. No one, I imagine, would dream of seeking for the cause of the development of the sixth finger and toe in the famous Maltese, in the direct influence of the external conditions of his life.

I conceive that both hereditary transmission and adaptation need to be analysed into their constituent conditions by the further application of the doctrine of the Struggle for Existence. It is a probable hypothesis, that what the world is to organisms in general, each organism is to the molecules of which it is composed. Multitudes of these, having diverse tendencies, are competing with one another for opportunity to exist and multiply; and the organism, as a whole, is as much the product of the molecules which are victorious as the Fauna, or Flora, of a country is the product of the victorious organic beings in it.

On this hypothesis, hereditary transmission is the result of the victory of particular molecules contained in the impregnated germ. Adaptation to conditions is the result of the favouring of the multiplication of those molecules whose organising tendencies are most in harmony with such conditions. In this view of the matter, conditions are not actively productive, but are passively permissive; they do not cause variation in any given direction, but they permit and favour a tendency in that direction which already exists.

It is true that, in the long run, the origin of the organic molecules themselves, and of their tendencies, is to be sought in the external world ; but if we carry our inquiries as far back as this, the distinction between internal and external impulses vanishes. On the other hand, if we confine ourselves to the consideration of a single organism, I think it must be admitted that the existence of an internal metamorphic tendency must be as distinctly recognised as that of an internal conservative tendency ; and that the influence of conditions is mainly, if not wholly, the result of the extent to which they favour the one, or the other, of these tendencies.

III. There is only one point upon which I fundamentally and entirely disagree with Professor Haeckel, but that is the very important one of his conception of geological time, and of the meaning of the stratified rocks as records and indications of that time. Conceiving that the stratified rocks of an epoch indicate a period of depression, and that the intervals between the epochs correspond with periods of elevation of which we have no record, he intercalates between the different epochs, or periods, intervals which he terms " Ante-periods." Thus, instead of considering the Triassic, Jurassic, Cretaceous, and Eocene periods, as continuously successive, he

interposes a period before each, as an " Antetrias-
zeit," " Antejura-zeit," " Antecreta-zeit," " Anteo-
cenzeit," &c.   And he conceives that the abrupt
changes between the Faunæ of the different formations are due to the lapse of time, of which we have
no organic record, during their " Ante-periods."

The frequent occurrence of strata containing
assemblages of organic forms which are intermediate between those of adjacent formations, is,
to my mind, fatal to this view.   In the well-
known St. Cassian beds, for example, Palæozoic
and Mesozoic forms are commingled, and, between
the Cretaceous and the Eocene formations, there
are similar transitional beds.   On the other hand,
in the middle of the Silurian series, extensive
unconformity of the strata indicates the lapse of
vast intervals of time between the deposit of
successive beds, without any corresponding change
in the Fauna.

Professor Haeckel will, I fear, think me unreasonable, if I say that he seems to be still overshadowed
by geological superstitions ; and that he will have
to believe in the completeness of the geological
record far less than he does at present.   He assumes,
for example, that there was no dry land, nor any
terrestrial life, before the end of the Silurian epoch,
simply because, up to the present time, no indications of fresh water, or terrestrial organisms, have
been found in rocks of older date.   And, in
speculating upon the origin of a given group, he

rarely goes further back than the " Ante-period," which precedes that in which the remains of animals belonging to that group are found. Thus, as fossil remains of the majority of the groups of *Reptilia* are first found in the Trias, they are assumed to have originated in the "Antetriassic" period, or between the Permian and Triassic epochs.

I confess this is wholly incredible to me. The Permian and the Triassic deposits pass completely into one another; there is no sort of discontinuity answering to an unrecorded "Antetrias"; and, what is more, we have evidence of immensely extensive dry land during the formation of these deposits. We know that the dry land of the Trias absolutely teemed with reptiles of all groups except Pterodactyles, Snakes, and perhaps Tortoises; there is every probability that true Birds existed, and *Mammalia* certainly did. Of the inhabitants of the Permian dry land, on the contrary, all that have left a record are a few lizards. Is it conceivable that these last should really represent the whole terrestrial population of that time, and that the development of Mammals, of Birds, and of the highest forms of Reptiles, should have been crowded into the time during which the Permian conditions quietly passed away, and the Triassic conditions began ? Does not any such supposition become in the highest degree improbable, when, in the terrestrial or fresh-water Labyrinthodonts, which lived on the land of the Carboniferous epoch, as well as on that of the Trias, we have evidence that one form of terrestrial life persisted, throughout all these ages, with no important modification ? For my part, having regard to the small amount of modification (except in the way of extinction) which the Crocodilian, Lacertilian, and Chelonian *Reptilia* have undergone, from the older Mesozoic times to the present day, I cannot but put the existence of the common stock from which they sprang far back in the Palæozoic epoch ; and I should apply a similar argumentation to all other groups of animals.

[The remainder of this essay contains a discussion of questions of taxonomy and phylogeny, which is now antiquated.   I have reprinted the considerations about the reconciliation of Teleology with Morphology, about " Dysteleology," and about the struggle for existence within the organism, because it has happened to me to be charged with overlooking them.

In discussing Teleology, I ought to have pointed out, as I have done elsewhere (*Life and Letters of Charles Darwin*, vol. ii. p. 202), that Paley "proleptically accepted the modern doctrine of Evolution," (*Natural Theology*, chap. xxiii.).   1893.]

V

MR. DARWIN'S CRITICS [1]

[1871]

THE gradual lapse of time has now separated us by more than a decade from the date of the publication of the "Origin of Species"— and whatever may be thought or said about Mr. Darwin's doctrines, or the manner in which he has propounded them, this much is certain, that, in a dozen years, the "Origin of Species" has worked as complete a revolution in biological science as the "Principia" did in astronomy—and it has done so, because, in the words of Helmholtz, it contains "an essentially new creative thought." [2]

And as time has slipped by, a happy change

[1] 1. *Contributions to the Theory of Natural Selection.* By A. R. Wallace. 1870.—2. *The Genesis of Species.* By St. George Mivart, F.R.S. Second Edition. 1871.—3. *Darwin's Descent of Man.* *Quarterly Review,* July 1871.

[2] Helmholtz: *Ueber das Ziel und die Fortschritte der Naturwissenschaft.* Eröffnungsrede für die Naturforscherversammlung zu Innsbruck. 1869.

has come over Mr. Darwin's critics. The mixture
of ignorance and insolence which, at first, characterised a large proportion of the attacks with which
he was assailed, is no longer the sad distinction of
anti-Darwinian criticism. Instead of abusive non-
sense, which merely discredited its writers, we read
essays, which are, at worst, more or less intelligent
and appreciative; while, sometimes, like that
which appeared in the " North British Review " for
1867, they have a real and permanent value.

The several publications of Mr. Wallace and Mr.
Mivart contain discussions of some of Mr. Darwin's
views, which are worthy of particular attention, not
only on account of the acknowledged scientific
competence of these writers, but because they exhibit an attention to those philosophical questions
which underlie all physical science, which is as rare
as it is needful. And the same may be said of an
article in the " Quarterly Review " for July 1871,
the comparison of which with an article in the
same Review for July 1860, is perhaps the best
evidence which can be brought forward of the
change which has taken place in public opinion
on " Darwinism."

The Quarterly Reviewer admits " the certainty
of the action of natural selection " (p. 49); and
further allows that there is an à priori probability
in favour of the evolution of man from some lower
animal form, if these lower animal forms themselves have arisen by evolution.

Mr. Wallace and Mr. Mivart go much further
than this. They are as stout believers in evolution
as Mr. Darwin himself; but Mr. Wallace denies
that man can have been evolved from a lower
animal by that process of nàtural selection which
he, with Mr. Darwin, holds to have been sufficient
for the evolution of all animals below man; while
Mr. Mivart, admitting that natural selection has
been one of the conditions of the evolution of the
animals below man, maintains that natural selection must, even in their case, have been supple-
mented by " some other cause "—of the nature of
which, unfortunately, he does not give us any idea.
Thus Mr. Mivart is less of a Darwinian than Mr.
Wallace, for he has less faith in the power of
natural selection. But he is more of an evolutionist
than Mr. Wallace, because Mr. Wallace thinks it
necessary to call in an intelligent agent—a sort of
supernatural Sir John Sebright—to produce even
the animal frame of man; while Mr. Mivart requires no Divine assistance till he comes to man's
soul.

Thus there is a considerable divergence between
Mr. Wallace and Mr. Mivart. On the other hand,
there are some curious similarities between Mr.
Mivart and the Quarterly Reviewer, and these
are sometimes so close, that, if Mr. Mivart thought
it worth while, I think he might make out a
good case of plagiarism against the Reviewer, who
studiously abstains from quoting him.

Both the Reviewer and Mr. Mivart reproach Mr. Darwin with being, "like so many other physicists," entangled in a radically false metaphysical system, and with setting at nought the first principles of both philosophy and religion. Both enlarge upon the necessity of a sound philosophical basis, and both, I venture to add, make a conspicuous exhibition of its absence. The Quarterly Reviewer believes that man "differs more from an elephant or a gorilla than do these from the dust of the earth on which they tread," and Mr. Mivart has expressed the opinion that there is more difference between man and an ape than there is between an ape and a piece of granite.[1]

And even when Mr. Mivart (p. 86) trips in a matter of anatomy, and creates a difficulty for Mr. Darwin out of a supposed close similarity between the eyes of fishes and cephalopods, which (as Gegenbaur and others have clearly shown) does not exist, the Quarterly Reviewer adopts the argument without hesitation (p. 66).

There is another important point, however, in which it is hard to say whether Mr. Mivart diverges from the Quarterly Reviewer or not.

The Reviewer declares that Mr. Darwin has, "with needless opposition, set at nought the first principles of both philosophy and religion" (p. 90).

---

[1] See the *Tablet* for March 11, 1871.

It looks, at first, as if this meant, that Mr.
Darwin's views being false, the opposition to
"religion" which flows from them must be needless.   But I suspect this is not the right view of
the meaning of the passage, as Mr. Mivart, from
whom the Quarterly Reviewer plainly draws so
much inspiration, tells us that "the consequences
which have been drawn from evolution, whether
exclusively Darwinian or not, to the prejudice of
religion, by no means follow from it, and are in
fact illegitimate" (p. 5).

I may assume, then, that the Quarterly
Reviewer and Mr. Mivart admit that there is no
necessary opposition between "evolution whether
exclusively Darwinian or not," and religion.   But
then, what do they mean by this last much-
abused term?   On this point the Quarterly
Reviewer is silent.   Mr. Mivart, on the contrary,
is perfectly explicit, and the whole tenor of his
remarks leaves no doubt that by "religion" he
means theology ; and by theology, that particular
variety of the great Proteus, which is expounded
by the doctors of the Roman Catholic Church, and
held by the members of that religious community
to be the sole form of absolute truth and of saving
faith.

According to Mr. Mivart, the greatest and most
orthodox authorities upon matters of Catholic
doctrine agree in distinctly asserting "derivative
creation" or evolution ; "and thus their teachings

harmonise with all that modern science can
possibly require " (p. 305).

I confess that this bold assertion interested me
more than anything else in Mr. Mivart's book.
What little knowledge I possessed of Catholic
doctrine, and of the influence exerted by Catholic
authority in former times, had not led me to
expect that modern science was likely to find
a warm welcome within the pale of the greatest
and most consistent of theological organisations.
And my astonishment reached its climax when
I found Mr. Mivart citing Father Suarez as his
chief witness in favour of the scientific freedom
enjoyed by Catholics—the popular repute of that
learned theologian and subtle casuist not being such
as to make his works a likely place of refuge for
liberality of thought. But in these days, when
Judas Iscariot and Robespierre, Henry VIII.
and Catiline, have all been shown to be men of
admirable virtue, far in advance of their age, and
consequently the victims of vulgar prejudice, it
was obviously possible that Jesuit Suarez might
be in like case. And, spurred by Mr. Mivart's
unhesitating declaration, I hastened to acquaint
myself with such of the works of the great Catholic
divine as bore upon the question, hoping, not
merely to acquaint myself with the true teachings
of the infallible Church, and free myself of an
unjust prejudice ; but, haply, to enable myself, at
a pinch, to put some Protestant bibliolater to

shame, by the bright example of Catholic freedom from the trammels of verbal inspiration.

I regret to say that my anticipations have been cruelly disappointed. But the extent to which my hopes have been crushed can only be fully appreciated by citing, in the first place, those passages of Mr. Mivart's work by which they were excited. In his introductory chapter I find the following passages :—

"The prevalence of this theory [of evolution] need alarm no one, for it is, without any doubt, perfectly consistent with the strictest and most orthodox Christian [1] theology " (p. 5).

"Mr. Darwin and others may perhaps be excused if they have not devoted much time to the study of Christian philosophy ; but they have no right to assume or accept without careful examination, as an unquestioned fact, that in that philosophy there is a necessary antagonism between the two ideas ' creation ' and ' evolution,' as applied to organic forms.

" It is notorious and patent to all who choose to seek, that many distinguished Christian thinkers have accepted, and do accept, both ideas, *i.e.* both ' creation ' and ' evolution.'

"As much as ten years ago an eminently Christian writer observed : ' The creationist theory does not necessitate the perpetual search after

---

[1] It should be observed that Mr. Mivart employs the term " Christian " as if it were the equivalent of " Catholic."

manifestations of miraculous power and perpetual
"catastrophes." Creation is not a miraculous
interference with the laws of Nature, but the very
institution of those laws. Law and regularity,
not arbitrary intervention, was the patristic ideal
of creation. With this notion they admitted,
without difficulty, the most surprising origin of
living creatures, provided it took place by *law*.
They held that when God said, " Let the waters
produce," " Let the earth produce," He conferred
forces on the elements of earth and water which
enabled them naturally to produce the various
species of organic beings. This power, they
thought, remains attached to the elements
throughout all time.' The same writer quotes
St. Augustin and St. Thomas Aquinas, to the
effect that, ' in the institution of Nature, we do not
look for miracles, but for the laws of Nature.'
And, again, St. Basil speaks of the continued
operation of natural laws in the production of all
organisms.

" So much for the writers of early and mediæval
times. As to the present day, the author can
confidently affirm that there are many as well
versed in theology as Mr. Darwin is in his own
department of natural knowledge, who would not
be disturbed by the thorough demonstration of his
theory. Nay, they would not even be in the least
painfully affected at witnessing the generation of
animals of complex organisation by the skilful

artificial arrangement of natural forces, and the production, in the future, of a fish by means analogous to those by which we now produce urea.

"And this because they know that the possibility of such phenomena, though by no means actually foreseen, has yet been fully provided for in the old philosophy centuries before Darwin, or even centuries before Bacon, and that their place in the system can be at once assigned them without even disturbing its order or marring its harmony.

"Moreover, the old tradition in this respect has never been abandoned, however much it may have been ignored or neglected by some modern writers. In proof of this, it may be observed that perhaps no post-mediæval theologian has a wider reception amongst Christians throughout the world than Suarez, who has a separate section [1] in opposition to those who maintain the distinct creation of the various kinds—or substantial forms—of organic life " (pp. 19—21).

Still more distinctly does Mr. Mivart express himself in the same sense, in his last chapter, entitled " Theology and Evolution " (pp. 302-5).

" It appears, then, that Christian thinkers are perfectly free to accept the general evolution theory. But are there any theological authorities to justify this view of the matter ?

[1] Suarez, *Metaphysica.* Edition Vivés. Paris, 1868, vol. i. Disput. xv. § 2.

"Now, considering how extremely recent are these biological speculations, it might hardly be expected à priori that writers of earlier ages should have given expression to doctrines harmonising in any degree with such very modern views; nevertheless, this is certainly the case, and it would be easy to give numerous examples. It will be better, however, to cite one or two authorities of weight. Perhaps no writer of the earlier Christian ages could be quoted whose authority is more generally recognised than that of St. Augustin. The same may be said of the mediæval period for St. Thomas Aquinas: and since the movement of Luther, Suarez may be taken as an authority, widely venerated, and one whose orthodoxy has never been questioned.

"It must be borne in mind that for a considerable time even after the last of these writers no one had disputed the generally received belief as to the small age of the world, or at least of the kinds of animals and plants inhabiting it. It becomes, therefore, much more striking if views formed under such a condition of opinion are found to harmonise with modern ideas concerning 'Creation' and organic Life.

"Now St. Augustin insists in a very remarkable manner on the merely derivative sense in which God's creation of organic forms is to be understood; that is, that God created them by conferring

on the material world the power to evolve them
under suitable conditions."

Mr. Mivart then cites certain passages from St.
Augustin, St. Thomas Aquinas, and Cornelius à
Lapide, and finally adds:—

"As to Suarez, it will be enough to refer to Disp. xv. sec. 2,
No. 9, p. 508, t. i. edition Vivés, Paris ; also Nos. 13—15.
Many other references to the same effect could easily be given,
but these may suffice.

"It is then evident that ancient and most venerable theological authorities distinctly assert *derivative* creation, and
thus their teachings harmonise with all that modern science
can possibly require."

It will be observed that Mr. Mivart refers solely
to Suarez's fifteenth Disputation, though he adds,
"Many other references to the same effect could
easily be given." I shall look anxiously for these
references in the third edition of the "Genesis of
Species." For the present, all I can say is, that
I have sought in vain, either in the fifteenth
Disputation, or elsewhere, for any passage in
Suarez's writings which, in the slightest degree,
bears out Mr. Mivart's views as to his opinions.[1]

The title of this fifteenth Disputation is "De
causa formali substantiali," and the second section
of that Disputation (to which Mr. Mivart refers)
is headed, "Quomodo possit forma substantialis
fieri in materia et ex materia?"

---

[1] The edition of Suarez's *Disputationes* from which the following citations are given, is Birckmann's, in two volumes folio,
and is dated 1630.

The problem which Suarez discusses in this place may be popularly stated thus : According to the scholastic philosophy every natural body has two components—the one its " matter " (*materia prima*), the other its " substantial form " (*forma substantialis*). Of these the matter is everywhere the same, the matter of one body being indistinguishable from the matter of any other body. That which differentiates any one natural body from all others is its substantial form, which inheres in the matter of that body, as the human soul inheres in the matter of the frame of man, and is the source of all the activities and other properties of the body.

Thus, says Suarez, if water is heated, and the source of heat is then removed, it cools again. The reason of this is that there is a certain " *intimius principium* " in the water, which brings it back to the cool condition when the external impediment to the existence of that condition is removed. This *intimius principium* is the " substantial form " of the water. And the substantial form of the water is not only the cause (*radix*) of the coolness of the water, but also of its moisture, of its density, and of all its other properties.

It will thus be seen that " substantial forms " play nearly the same part in the scholastic philosophy as " forces " do in modern science ; the general tendency of modern thought being to conceive all bodies as resolvable into material

particles and forces, in virtue of which last these
particles assume those dispositions and exercise
those powers which are characteristic of each
particular kind of matter.

But the Schoolmen distinguished two kinds of
substantial forms, the one spiritual and the other
material. The former division is represented by
the human soul, the *anima rationalis;* and they
affirm as a matter, not merely of reason, but of
faith, that every human soul is created out of
nothing, and by this act of creation is endowed
with the power of existing for all eternity, apart
from the *materia prima* of which the corporeal
frame of man is composed. And the *anima
rationalis,* once united with the *materia prima* of
the body, becomes its substantial form, and is the
source of all the powers and faculties of man—of
all the vital and sensitive phenomena which he
exhibits—just as the substantial form of water is
the source of all its qualities.

The "material substantial forms" are those
which inform all other natural bodies except that
of man; and the object of Suarez in the present
Disputation, is to show that the axiom " *ex nihilo
nihil fit,*" though not true of the substantial form
of man, is true of the substantial forms of all
other bodies, the endless mutations of which
constitute the ordinary course of nature. The
origin of the difficulty which he discusses is easily
comprehensible. Suppose a piece of bright iron

"13. Secundo de omnibus aliis formis substantialibus [sc. materialibus] dicendum est non fieri proprie ex nihilo, sed ex potentia præjacentis materiæ educi : ideoque in effectione harum formarum nil fieri contra illud axioma, *Ex nihilo nihil fit*, si recte intelligatur.   Hæc assertio sumitur ex Aristotele 1. Physicorum per totum et libro 7. Metaphyss. et ex aliis auctoribus, quos statim referam. Et declaratur breviter, nam fieri ex nihilo duo dicit, unum est fieri absolute et simpliciter, aliud est quod talis effectio fit ex nihilo. Primum propriè dicitur de re subsistente, quia ejus est fieri, cujus est esse : id autem proprie quod subsistit et habet esse ; nam quod alteri adjacet, potius est quo aliud est. Ex hac ergo parte, formæ substantiales materiales non fiunt ex nihilo, quia proprie non fiunt. Atque hanc rationem reddit Divus Thomas 1 parte, quæstione 45, articulo 8, et quæstione 90, articulo 2, et ex dicendis magis explicabitur. Sumendo ergo ipsum *fieri* in hac proprietate et rigore, sic fieri ex nihilo est fieri secundum se totum, id est nulla sui parte præsupposita, ex quo fiat. Et hac ratione res naturales dum de novo fiunt, non fiunt ex nihilo, quia fiunt ex præsupposita materia, ex qua componuntur, et ita non fiunt, secundum se totæ, sed secundum aliquid sui. Formæ autem harum rerum, quamvis revera totam suam entitatem de novo accipiant, quam antea non habebant, quia vero ipsæ non fiunt, ut dictum est, ideo neque ex nihilo fiunt. Attamen, quia latiori modo sumendo verbum illud *fieri* negari non potest : quin forma facta sit, eo modo quo nunc est, et antea non erat, ut etiam probat ratio dubitandi posita in principio sectionis, ideo addendum est, sumpto *fieri* in hac amplitudine, fieri ex nihilo non tamen negare habitudinem materialis causæ intrinsecè componentis id quod fit, sed etiam habitudinem causæ materialis per se causantis et sustentantis formam quæ fit, seu confit. Diximus enim in superioribus materiam et esse causam compositi et formæ dependentis ab illa : ut res ergo dicatur ex nihilo fieri uterque modus causalitatis negari debet ; et eodem sensu accipiendum est illud axioma, ut sit verum : *Ex nihilo nihil fit*, scilicet virtute agentis naturalis et finiti nihil fieri, nisi ex præsupposito subjecto per se concurrente, et ad compositum et ad formam, si utrumque suo modo ab eodem agente fiat. Ex his ergo rectè concluditur, formas substantiales materiales non fieri ex nihilo, quia fiunt ex materia, quæ in suo genere per se concurrit, et influit ad esse, et fieri talium formarum ; quia, sicut esse non possunt nisi affixæ materiæ, a qua sustententur in esse : ita nec fieri possunt, nisi earum effectio et penetratio in eadem materia sustentetur. Et hæc est propria et per se differentia inter effectionem ex nihilo, et ex aliquo, propter quam, ut infra ostendemus, prior modus efficiendi superat vim finitam naturaliam agentium, non vero posterior.

"14. Ex his etiam constat, proprie de his formis dici non creari, sed educi de potentia materiæ." [1]

If I may venture to interpret these hard sayings, Suarez conceives that the evolution of substantial forms in the ordinary course of nature, is conditioned not only by the existence of the *materia prima*, but also by a certain " concurrence and influence " which that *materia* exerts ; and every new substantial form being thus conditioned, and in part, at any rate, caused, by a pre-existing something, cannot be said to be created out of nothing.

But as the whole tenor of the context shows, Suarez applies this argumentation merely to the evolution of material substantial forms in the ordinary course of nature. How the substantial forms of animals and plants primarily originated, is a question to which, so far as I am able to discover, he does not so much as allude in his "Metaphysical Disputations." Nor was there any necessity that he should do so, inasmuch as he

---

[1] Suarez, *loc. cit.* Disput. xv. § ii.

has devoted a separate treatise of considerable
bulk to the discussion of all the problems which
arise out of the account of the Creation which is
given in the Book of Genesis. And it is a
matter of wonderment to me that Mr. Mivart,
who somewhat sharply reproves "Mr. Darwin and
others" for not acquainting themselves with the
true teachings of his Church, should allow
himself to be indebted to a heretic like myself
for a knowledge of the existence of that "Tractatus de opere sex Dierum,"[1] in which the learned
Father, of whom he justly speaks, as "an
authority widely venerated, and whose orthodoxy
has never been questioned," directly opposes all
those opinions for which Mr. Mivart claims the
shelter of his authority.

In the tenth and eleventh chapters of the first
book of this treatise, Suarez inquires in what sense
the word "day," as employed in the first chapter
of Genesis, is to be taken. He discusses the
views of Philo and of Augustin on this question,
and rejects them. He suggests that the approval
of their allegorising interpretations by St. Thomas
Aquinas, merely arose out of St. Thomas's
modesty, and his desire not to seem openly to
controvert St. Augustin—"voluisse Divus Thomas

---

[1] *Tractatus de opere sex Dierum, seu de Universi Creatione,
quatenus sex diebus perfecta esse, in libro Genesis cap. i. refertur,
et præsertim de productione hominis in statu innocentiæ.* Ed.
Birckmann, 1622.

pro sua modestia subterfugere vim argumenti
potius quam aperte Augustinum inconstantiæ
arguere."

Finally, Suarez decides that the writer of
Genesis meant that the term "day" should be
taken in its natural sense; and he winds up the
discussion with the very just and natural remark
that "it is not probable that God, in inspiring
Moses to write a history of the Creation which
was to be believed by ordinary people, would
have made him use language, the true meaning of
which it is hard to discover, and still harder to
believe." [1]

And in chapter xii. 3, Suarez further observes :—

"Ratio enim retinendi veram significationem diei naturalis
est illa communis, quod verba Scripturæ non sunt ad metaphoras
transferenda, nisi vel necessitas cogit, vel ex ipsa scriptura
constet, et maximè in historica narratione et ad instructionem
fidei pertinente : sed hæc ratio non minus cogit ad intelligendum
propriè dierum numerum, quam diei qualitatem, QUIA NON
MINUS UNO MODO QUAM ALIO DESTRUITUR SINCERITAS, IMO ET
VERITAS HISTORIÆ. Secundo hoc valde confirmant alia Scripturæ
loca, in quibus hi sex dies tanquam veri, et inter se distincti
commemorantur, ut Exod. 20 dicitur, *Sex diebus operabis et
facies omnia opera tua, septimo autem die Sabbatum Domini Dei*

---

[1] "Propter hæc ergo sententia illa Augustini et propter nimiam
obscuritatem et subtilitatem ejus difficilis creditu est : quia
verisimile non est Deum inspirasse Moysi, ut historiam de
creatione mundi ad fidem totius populi adeo necessariam per
nomina dierum explicaret, quorum significatio vix inveniri et
difficillime ab aliquo credi posset." (*Loc. cit.* Lib. I. cap. xi.
42.)

tui est. Et infra : *Sex enim diebus fecit Dominus cœlum et terram et mare et omnia quœ in eis sunt,* et idem repetitur in cap. 31. In quibus locis sermonis proprietas colligi potest tum ex æquiparatione, nam cum dicitur : *sex diebus operabis,* propriissimè intelligitur : tum quia non est verisimile, potuisse populum intelligere verba illa in alio sensu, et è contrario incredibile est, Deum in suis præceptis tradendis illis verbis ad populum fuisse loquutum, quibus deciperetur, falsum sensum concipiendo, si Deus non per sex veros dies opera sua fecisset."

These passages leave no doubt that this great doctor of the Catholic Church, of unchallenged authority and unspotted orthodoxy, not only declares it to be Catholic doctrine that the work of creation took place in the space of six natural days ; but that he warmly repudiates, as inconsistent with our knowledge of the Divine attributes, the supposition that the language which Catholic faith requires the believer to hold that God inspired, was used in any other sense than that which He knew it would convey to the minds of those to whom it was addressed.

And I think that in this repudiation Father Suarez will have the sympathy of every man of common uprightness, to whom it is certainly "incredible" that the Almighty should have acted in a manner which He would esteem dishonest and base in a man.

But the belief that the universe was created in six natural days is hopelessly inconsistent with the doctrine of evolution, in so far as it applies to the stars and planetary bodies ; and it can be

made to agree with a belief in the evolution of
living beings only by the supposition that the
plants and animals, which are said to have been
created on the third, fifth, and sixth days, were
merely the primordial forms, or rudiments, out of
which existing plants and animals have been
evolved; so that, on these days, plants and
animals were not created actually, but only
potentially.

The latter view is that held by Mr. Mivart, who
follows St. Augustin, and implies that he has the
sanction of Suarez. But, in point of fact, the
latter great light of orthodoxy takes no small
pains to give the most explicit and direct contradiction to all such imaginations, as the following
passages prove. In the first place, as regards
plants, Suarez discusses the problem :—

"*Quomodo herba virens et cætera vegetabilia hoc [tertio] die
fuerint producta.*[1]

" Præcipua enim difficultas hîc est, quam attingit Div. Thomas
1, par. qu. 69, art. 2, an hæc productio plantarum hoc die facta
intelligenda sit de productione ipsarum in proprio esse actuali et
formali (ut sic rem explicerem) vel de productione tantum in
semine et in potentia. Nam Divus Augustinus libro quinto Genes.
ad liter. cap. 4 et 5 et libro 8, cap. 3, posteriorem partem tradit,
dicens, terram in hoc die accepisse virtutem germinandi omnia
vegetabilia quasi concepto omnium illorum semine, non tamen
statim vegetabilia omnia produxisse. Quod primo suadet verbis
illis capitis secundi. *In die quo fecit Deus cælum et terram et*

---

[1] *Loc. cit.* Lib. II. cap. vii. et viii. 1, 32, 35.

omne virgultum agri priusquam germinaret. Quomodo enim
potuerunt virgulta fieri antequam terra germinaret nisi quia
causaliter prius et quasi in radice, seu in semine facta sunt, et
postea in actu producta? Secundo confirmari potest, quia
verbum illud *germinet terra* optime exponitur potestative ut sic
dicam, id est accipiat terra vim germinandi. Sicut in eodem
capite dicitur *crescite et multiplicamini.* Tertio potest confirmari,
quia actualis productio vegetabilium non tam ad opus creationis,
quam ad opus propagationis pertinet, quod postea factum est.
Et hanc sententiam sequitur Eucherius lib. 1, in Gen. cap. 11, et
illi faveat Glossa, interli. Hugo. et Lyran. dum verbum
germinet dicto modo exponunt. NIHILOMINUS CONTRARIA
SENTENTIA TENENDA EST : SCILICET, PRODUXISSE DEUM HOC
DIE HERBAM, ARBORES, ET ALIA VEGETABILIA ACTU IN PROPRIA
SPECIE ET NATURA. Hæc est communis sententia Patrum.—
Basil. homil. 5 ; Exæmer. Ambros. lib. 3 ; Exæmer. cap. 8,
11, et 16 ; Chrysost. homil. 5 in Gen. Damascene. lib. 2 de Fid.
cap. 10 ; Theodor. Cyrilli. Bedæ, Glossæ ordinariæ et aliorum in
Gen. Et idem sentit Divus Thomas, *supra*, solvens argumenta
Augustini, quamvis propter reverentiam ejus quasi problematice
semper procedat. Denique idem sentiunt omnes qui in his
operibus veram successionem et temporalem distinctionem
agnoscant."

Secondly, with respect to animals, Suarez is no
less decided :—

" *De animalium ratione carentium productione quinto et sexto
die facta.*[1]

"32. Primo ergo nobis certum sit hæc animantia non in
virtute tantum aut in semine, sed actu, et in seipsis, facta fuisse
his diebus in quibus facta narrantur. Quanquam Augustinus
lib. 3, Gen. ad liter. cap. 5 in sua persistens sententia contrarium
sentire videatur."

But Suarez proceeds to refute Augustin's

---

[1] *Loc. cit.* Lib. II. cap. vii. et viii. 1, 32, 35.

opinions at great length, and his final judgment
may be gathered from the following passage :—

"35. Tertio dicendum est, hæc animalia omnia his diebus
producta esse, IN PERFECTO STATU, IN SINGULIS INDIVIDUIS, SEU
SPECIEBUS SUIS, JUXTA UNIUSCUJUSQUE NATURAM . . . .
ITAQUE FUERUNT OMNIA CREATA INTEGRA ET OMNIBUS SUIS
MEMBRIS PERFECTA."

As regards the creation of animals and plants,
therefore, it is clear that Suarez, so far from
"distinctly asserting derivative creating," denies
it as distinctly and positively as he can ; that
he is at much pains to refute St. Augustin's
opinions ; that he does not hesitate to regard
the faint acquiescence of St. Thomas Aquinas in
the views of his brother saint as a kindly subter-
fuge on the part of Divus Thomas ; and that he
affirms his own view to be that which is supported
by the authority of the Fathers of the Church.
So that, when Mr. Mivart tells us that Catholic
theology is in harmony with all that modern
science can possibly require ; that " to the general
theory of evolution, and to the special Darwinian
form of it, no exception . . . need be taken on
the ground of orthodoxy ;" and that "law and
regularity, not arbitrary intervention, was the
Patristic ideal of creation," we have to choose
between his dictum, as a theologian, and that
of a great light of his Church, whom he himself declares to be "widely venerated as an

authority, and whose orthodoxy has never been questioned."

But Mr. Mivart does not hesitate to push his attempt to harmonise science with Catholic orthodoxy to its utmost limit ; and, while assuming that the soul of man "arises from immediate and direct creation," he supposes that his body was "formed at first (as now in each separate individual) by derivative, or secondary creation, through natural laws " (p. 331).

This means, I presume, that an animal, having the corporeal form and bodily powers of man, may have been developed out of some lower form of life by a process of evolution; and that, after this anthropoid animal had existed for a longer or shorter time, God made a soul by direct creation, and put it into the manlike body, which, heretofore, had been devoid of that *anima rationalis*, which is supposed to be man's distinctive character.

This hypothesis is incapable of either proof or disproof, and therefore may be true ; but if Suarez is any authority, it is not Catholic doctrine. "Nulla est in homine forma educta de potentia materiæ," [1] is a dictum which is absolutely inconsistent with the doctrine of the natural evolution of any vital manifestation of the human body.

Moreover, if man existed as an animal before

[1] Disput. xv. § x. No. 27

he was provided with a rational soul, he must, in accordance with the elementary requirements of the philosophy in which Mr. Mivart delights, have possessed a distinct sensitive and vegetative soul, or souls. Hence, when the "breath of life" was breathed into the manlike animal's nostrils, he must have already been a living and feeling creature. But Suarez particularly discusses this point, and not only rejects Mr. Mivart's view, but adopts language of very theological strength regarding it.

"Possent præterea his adjungi argumenta theologica, ut est illud quod sumitur ex illis verbis Genes. 2. *Formavit Deus hominem ex limo terræ et inspiravit in faciem ejus spiraculum vitæ et factus est homo in animam viventem :* ille enim spiritus, quam Deus spiravit, anima rationalis fuit, et PER EADEM FACTUS EST HOMO VIVENS, ET CONSQUENTER, ETIAM SENTIENS.

"Aliud est ex VIII. Synodo Generali quæ est Constantinopolitana IV. can. 11, qui sic habet. *Apparet quosdam in tantum impietatis venisse ut homines duas animas habere dogmatizent : talis igitur impietatis inventores et similes sapientes, cum Vetus et Novum Testamentum omnesque Ecclesiæ patres unam animam rationalem hominem habere asseverent, Sancta et universalis Synodus anathematizat.*" [1]

Moreover, if the animal nature of man was the result of evolution, so must that of woman have been. But the Catholic doctrine, according to Suarez, is that woman was, in the strictest and most literal sense of the words, made out of the rib of man.

[1] Disput. xv. "De causa formali substantiali," § x. No. 24.

"Nihilominus sententia Catholica est, verba illa Scripturæ esse ad literam intelligenda. AC PROINDE VERE, AC REALITER, TULISSE DEUM COSTAM ADAMÆ, ET, EX ILLA, CORPUS EVÆ FORMASSE." [1]

Nor is there any escape in the supposition that some woman existed before Eve, after the fashion of the Lilith of the rabbis; since Suarez qualifies that notion, along with some other Judaic imaginations, as simply "damnabilis." [2] After the perusal of the "Tractatus de Opere" it is, in fact, impossible to admit that Suarez held any opinion respecting the origin of species, except such as is consistent with the strictest and most literal interpretation of the words of Genesis. For Suarez, it is Catholic doctrine, that the world was made in six natural days. On the first of these days the *materia prima* was made out of nothing, to receive afterwards those "substantial forms" which moulded it into the universe of things; on the third day, the ancestors of all living plants suddenly came into being, full-grown, perfect, and possessed of all the properties which now distinguish them; while, on the fifth and sixth days, the ancestors of all existing animals were similarly caused to exist in their complete and perfect state, by the infusion of their appropriate material substantial forms into the matter

---

[1] *Tractatus de Opere*, Lib. III. "De hominis creatione," cap. ii. No. 3.

[2] *Ibid.* Lib. III. cap. iv. Nos. 8 and 9

which had already been created. Finally, on the
sixth day, the *anima rationalis*—that rational and
immortal substantial form which is peculiar to
man—was created out of nothing, and " breathed
into " a mass of matter which, till then, was mere
dust of the earth, and so man arose. But the
species man was represented by a solitary male
individual, until the Creator took out one of his
ribs and fashioned it into a female.

This is the view of the " Genesis of Species "
held by Suarez to be the only one consistent with
Catholic faith : it is because he holds this view to
be Catholic that he does not hesitate to declare
St. Augustin unsound, and St. Thomas Aquinas
guilty of weakness, when the one swerved from
this view and the other tolerated the deviation.
And, until responsible Catholic authority—say,
for example, the Archbishop of Westminster—
formally declares that Suarez was wrong, and
that Catholic priests are free to teach their flocks
that the world was *not* made in six natural days,
and that plants and animals were *not* created in
their perfect and complete state, but have been
evolved by natural processes through long ages
from certain germs in which they were potentially
contained, I, for one, shall feel bound to believe
that the doctrines of Suarez are the only ones
which are sanctioned by Infallible Authority, as
represented by the Holy Father and the Catholic
Church.

I need hardly add that they are as absolutely denied and repudiated by Scientific Authority, as represented by Reason and Fact. The question whether the earth and the immediate progenitors of its present living population were made in six natural days or not is no longer one upon which two opinions can be held.

The fact that it did not so come into being stands upon as sound a basis as any fact of history whatever. It is not true that existing plants and animals came into being within three days of the creation of the earth out of nothing, for it is certain that innumerable generations of other plants and animals lived upon the earth before its present population. And when, Sunday after Sunday, men who profess to be our instructors in righteousness read out the statement "In six days the Lord made heaven and earth, the sea, and all that in them is," in innumerable churches, they are either propagating what they may easily know, and, therefore, are bound to know, to be falsities ; or, if they use the words in some non-natural sense, they fall below the moral standard of the much-abused Jesuit.

Thus far the contradiction between Catholic verity and Scientific verity is complete and absolute, quite independently of the truth or false-hood of the doctrine of evolution. But, for those who hold the doctrine of evolution, all the Catholic verities about the creation of living beings must

be no less false. For them, the assertion that the
progenitors of all existing plants were made on the
third day, of animals on the fifth and sixth days,
in the forms they now present, is simply false.
Nor can they admit that man was made suddenly
out of the dust of the earth ; while it would be an
insult to ask an evolutionist whether he credits the
preposterous fable respecting the fabrication of
woman to which Suarez pins his faith. If Suarez
has rightly stated Catholic doctrine, then is
evolution utter heresy. And such I believe it to
be. In addition to the truth of the doctrine of
evolution, indeed, one of its greatest merits in
my eyes, is the fact that it occupies a position of
complete and irreconcilable antagonism to that
vigorous and consistent enemy of the highest intellectual, moral, and social life of mankind—the
Catholic Church. No doubt, Mr. Mivart, like
other putters of new wine into old bottles, is
actuated by motives which are worthy of respect,
and even of sympathy ; but his attempt has met
with the fate which the Scripture prophesies for
all such.

Catholic theology, like all theologies which are
based upon the assumption of the truth of the
account of the origin of things given in the Book
of Genesis, being utterly irreconcilable with the
doctrine of evolution, the student of science, who is
satisfied that the evidence upon which the doctrine
of evolution rests, is incomparably stronger and

better than that upon which the supposed authority of the Book of Genesis rests, will not trouble
himself further with these theologies, but will
confine his attention to such arguments against
the view he holds as are based upon purely
scientific data—and by .scientific data I do not
merely mean the truths of physical, mathematical,
or logical science, but those of moral and metaphysical science. For by science I understand
all knowledge which rests upon evidence and
reasoning of a like character to that which claims
our assent to ordinary scientific propositions. And
if any one is able to make good the assertion that
his theology rests upon valid evidence and sound
reasoning, then it appears to me that such theology
will take its place as a part of science.

The present antagonism between theology and
science does not arise from any assumption by the
men of science that all theology must necessarily
be excluded from science, but simply because
they are unable to allow that reason and morality
have two weights and two measures ; and that the
belief in a proposition, because authority tells you
it is true, or because you wish to believe it, which
is a high crime and misdemeanour when the subject matter of reasoning is of one kind, becomes
under the *alias* of "faith" the greatest of all
virtues when the subject matter of reasoning is of
another kind.

The Bishop of Brechin said well the other

day :—" Liberality in religion— I do not mean
tender and generous allowances for the mistakes of others—is only unfaithfulness to truth." [1]
And, with the same qualification, I venture
to paraphrase the Bishop's dictum : " Eccle-
siasticism in science is only unfaithfulness to
truth."

Elijah s great question, " Will you serve God or
Baal ? Choose ye," is uttered audibly enough in
the ears of every one of us as we come to man-
hood. Let every man who tries to answer it
seriously ask himself whether he can be satisfied
with the Baal of authority, and with all the good
things his worshippers are promised in this world
and the next. If he can, let him, if he be so
inclined, amuse himself with such scientific imple-
ments as authority tells him are safe and will not
cut his fingers ; but let him not imagine he is, or
can be, both a true son of the Church and a loyal
soldier of science.

And, on the other hand, if the blind acceptance
of authority appears to him in its true colours, as
mere private judgment *in excelsis*, and if he have
the courage to stand alone, face to face with the
abyss of the eternal and unknowable, let him be
content, once for all, not only to renounce the good
things promised by " Infallibility," but even to
bear the bad things which it prophesies ; content

[1] Charge at the Diocesan Synod of Brechin. *Scotsman*, Sept.
14, 1871.

to follow reason and fact in singleness and honesty
of purpose, wherever they may lead, in the sure
faith that a hell of honest men will, to him, be
more endurable than a paradise full of angelic
shams.

Mr. Mivart asserts that " without a belief in a
personal God there is no religion worthy of the
name." This is a matter of opinion. But it may
be asserted, with less reason to fear contradiction,
that the worship of a personal God, who, on Mr.
Mivart's hypothesis, must have used language
studiously calculated to deceive His creatures and
worshippers, is " no religion worthy of the name."
" Incredible est, Deum illis verbis ad populum
fuisse locutum quibus deciperetur," is a verdict in
which, for once, Jesuit casuistry concurs with the
healthy moral sense of all mankind.

Having happily got quit of the theological
aspect of evolution, the supporter of that great
truth who turns to the scientific objections which
are brought against it by recent criticism, finds, to
his relief, that the work before him is greatly
lightened by the spontaneous retreat of the enemy
from nine-tenths of the territory ·which he occupied ten years ago. Even the Quarterly Reviewer
not only abstains from venturing to deny that
evolution has taken place, but he openly admits
that Mr. Darwin has forced on men's minds " a
recognition of the probability, if not more, of

evolution, and of the certainty of the action of natural selection " (p. 49).

I do not quite see, myself, how, if the action of natural selection is *certain*, the occurrence of evolution is only *probable;* inasmuch as the development of a new species by natural selection is, so far as it goes, evolution. However, it is not worth while to quarrel with the precise terms of a sentence which shows that the high water mark of intelligence among those most respectable of Britons, the readers of the *Quarterly Review*, has now reached such a level that the next tide may lift them easily and pleasantly on the once-dreaded shore of evolution. Nor, having got there, do they seem likely to stop, until they have reached the inmost heart of that great region, and accepted the ape ancestry of, at any rate, the body of man. For the Reviewer admits that Mr. Darwin can be said to have established :

"That if the various kinds of lower animals have been evolved one from the other by a process of natural generation or evolution, then it becomes highly probable, *à priori*, that man's body has been similarly evolved ; but this, in such a case, becomes equally probable from the admitted fact that he is an animal at all " (p. 65).

From the principles laid down in the last sentence it would follow that if man were constructed upon a plan as different from that of any other animal as that of a sea-urchin is from that of a whale, it would be " equally probable " that he

had been developed from some other animal as it
is now, when we know that for every bone, muscle,
tooth, and even pattern of tooth, in man, there is a
corresponding bone, muscle, tooth, and pattern of
tooth, in an ape. And this shows one of two things
—either that the Quarterly Reviewer's notions of
probability are peculiar to himself, or that he has
such an overpowering faith in the truth of evolution
that no extent of structural break between one
animal and another is sufficient to destroy his conviction that evolution has taken place.

But this by the way. The importance of the
admission that there is nothing in man's physical
structure to interfere with his having been evolved
from an ape is not lessened because it is grudg-
ingly made and inconsistently qualified. And instead of jubilating over the extent of the enemy's
retreat, it will be more worth while to lay siege to
his last stronghold—the position that there is a
distinction in kind between the mental faculties
of man and those of brutes, and that in consequence
of this distinction in kind no gradual progress
from the mental faculties of the one to those of the
other can have taken place.

The Quarterly Reviewer entrenches himself
within formidable-looking psychological outworks,
and there is no getting at him without attacking
them one by one.

He begins by laying down the following proposition. " 'Sensation' is not 'thought,' and no

amount of the former would constitute the most rudimentary condition of the latter, though sensations supply the conditions for the existence of 'thought' or 'knowledge' " (p. 67).

This proposition is true, or not, according to the sense in which the word "thought" is employed. Thought is not uncommonly used in a sense co-extensive with consciousness, and, especially, with those states of consciousness we call memory. If I recall the impression made by a colour or an odour, and distinctly remember blueness or muskiness, I may say with perfect propriety that I "think of " blue or musk; and, so long as the thought lasts, it is simply a faint reproduction of the state of consciousness to which I gave the name in question, when it first became known to me as a sensation.

Now, if that faint reproduction of a sensation, which we call the memory of it, is properly termed a thought, it seems to me to be a somewhat forced proceeding to draw a hard and fast line of demarcation between thoughts and sensations. If sensations are not rudimentary thoughts, it may be said that some thoughts are rudimentary sensations. No amount of sound constitutes an echo, but for all that no one would pretend that an echo is something of totally different nature from a sound. Again, nothing can be looser, or more inaccurate, than the assertion that "sensations supply the conditions for the existence of thought or knowledge." If this implies that sensations supply the conditions for the existence of our memory of sensations or of our thoughts about sensations, it is a truism which it is hardly worth while to state so solemnly.   If it implies that sensations supply anything else, it is obviously erroneous. And if it means, as the context would seem to show it does, that sensations are the subject-matter of all thought or knowledge, then it is no less contrary to fact, inasmuch as our emotions, which constitute a large part of the subject-matter of thought or of knowledge, are not sensations.

More eccentric still is the Quarterly Reviewer's next piece of psychology.

" Altogether, we may clearly distinguish at least six kinds of action to which the nervous system ministers :—

" I. That in which impressions received result in appropriate movements without the intervention of sensation or thought, as in the cases of injury above given.—This is the reflex action of the nervous system.

" II. That in which stimuli from without result in sensations through the agency of which their due effects are wrought out. —Sensation.

" III. That in which impressions received result in sensations which give rise to the observation of sensible objects.—Sensible perception.

"IV. That in which sensations and perceptions continue to coalesce, agglutinate, and combine in more or less complex aggregations, according to the laws of the association of sensible perceptions.—Association.

" The above four groups contain only indeliberate operations, consisting, as they do at the best, but of mere *presentative* sensible ideas in no way implying any reflective or *representative* faculty. Such actions minister to and form *Instinct.* Besides these, we may distinguish two other kinds of mental action, namely :—

"V. That in which sensations and sensible perceptions are reflected on by thought, and recognised as our own, and wo ourselves recognised by ourselves as affected and perceiving.— Self-consciousness.

"VI. That in which we reflect upon our sensations or perceptions, and ask what they are, and why they are.—Reason.

"These two latter kinds of action are deliberate operations, performed, as they are, by means of representative ideas implying the use of a *reflective representative* faculty. Such actions distinguish the *intellect* or rational faculty. Now, we assert that possession in perfection of all the first four (*presentative*) kinds of action by no means implies the possession of the last two (*representative*) kinds. All persons, we think, must admit the truth of the following proposition :—

"Two faculties are distinct, not in degree but *in kind,* if we may possess the one in perfection without that fact implying that we possess the other also. Still more will this be the case if the two faculties tend to increase in an inverse ratio. Yet this is the distinction between the *instinctive* and the *intellectual* parts of man's nature.

"As to animals, we fully admit that they may possess all the first four groups of actions—that they may have, so to speak, mental images of sensible objects combined in all degrees of complexity, as governed by the laws of association. We deny to them, on the other hand, the possession of the last two kinds of mental action. We deny them, that is, the power of reflecting on their own existences, or of inquiring into the nature of objects and their causes. We deny that they know that they know or know themselves in knowing. In other words, we deny them *reason.* The possession of the presentative faculty, as above explained, in no way implies that of the reflective faculty ; nor does any amount of direct operation imply the power of asking the reflective question before mentioned, as to ' what' and ' why.' " (*Loc. cit.* pp. 67, 68.)

Sundry points are worthy of notice in this remarkable account of the intellectual powers. In the first place the Reviewer ignores emotion and

volition, though they are no inconsiderable "kinds
of action to which the nervous system ministers,"
and memory has a place in his classification only
by implication. Secondly, we are told that the
second "kind of action to which the nervous
system ministers" is "that in which stimuli from
without result in sensations through the agency
of which their due effects are wrought out.—
Sensation." Does this really mean that, in the
writer's opinion, "sensation" is the "agent" by
which the "due effect" of the stimulus, which
gives rise to sensation, is "wrought out"?
Suppose somebody runs a pin into me. The
"due effect" of that particular stimulus will
probably be threefold; namely, a sensation of
pain, a start, and an interjectional expletive.
Does the Quarterly Reviewer really think that
the "sensation" is the "agent" by which the
other two phenomena are wrought out?

But these matters are of little moment to
anyone but the Reviewer and those persons who
may incautiously take their physiology, or psychology, from him. The really interesting point is
this, that when he fully admits that animals
"may possess all the first four groups of actions,"
he grants all that is necessary for the purposes of
the evolutionist. For he hereby admits that in
animals "impressions received result in sensations
which give rise to the observation of sensible
objects," and that they have what he calls

"sensible perception." Nor was it possible to
help the admission ; for we have as much reason
to ascribe to animals, as we have to attribute to
our fellow-men, the power, not only of perceiving
external objects as external, and thus practically
recognizing the difference between the self and the
not-self; but that of distinguishing between like
and unlike, and between simultaneous and successive things. When a gamekeeper goes out
coursing with a greyhound in leash, and a hare
crosses the field of vision, he becomes the subject
of those states of consciousness we call visual
sensation, and that is all he receives from without.
Sensation, as such, tells him nothing whatever
about the cause of these states of consciousness;
but the thinking faculty instantly goes to work
upon the raw material of sensation furnished to it
through the eye, and gives rise to a train of
thoughts. First comes the thought that there is
an object at a certain distance ; then arises
another thought—the perception of the likeness
between the states of consciousness awakened by
this object to those presented by memory, as, on
some former occasion, called up by a hare ; this is
succeeded by another thought of the nature of an
emotion—namely, the desire to possess the hare ;
then follows a longer or shorter train of other
thoughts, which end in a volition and an act—the
loosing of the greyhound from the leash. These
several thoughts are the concomitants of a process

psychosis vanish, and the loosing of the dog follows
unconsciously, or as we say, without thinking about
it, upon the sight of the hare. No one will deny
that the series of acts which originally intervened
between the sensation and the letting go of the
dog were, in the strictest sense, intellectual and
rational operations. Do they cease to be so when
the man ceases to be conscious of them ? That
depends upon what is the essence and what the
accident of those operations, which, taken together, constitute ratiocination.

Now ratiocination is resolvable into predication,
and predication consists in marking, in some way,
the existence, the co-existence, the succession, the
likeness and unlikeness, of things or their ideas.
Whatever does this, reasons ; and if a machine produces the effects of reason, I see no more ground
for denying to it the reasoning power, because it
is unconscious, than I see for refusing to Mr.
Babbage's engine the title of a calculating machine
on the same grounds.

Thus it seems to me that a gamekeeper reasons,
whether he is conscious or unconscious, whether
his reasoning is carried on by neurosis alone, or
whether it involves more or less psychosis. And
if this is true of the gamekeeper, it is also true of
the greyhound. The essential resemblances in all
points of structure and function, so far as they can
be studied, between the nervous system of the man
and that of the dog, leave no reasonable doubt

that the processes which go on in the one are just like those which take place in the other. In the dog, there can be no doubt that the nervous matter which lies between the retina and the muscles undergoes a series of changes, precisely analogous to those which, in the man, give rise to sensation, a train of thought, and volition.

Whether this neurosis is accompanied by such psychosis as ours it is impossible to say ; but those who deny that the nervous changes, which, in the dog, correspond with those which underlie thought in a man, are accompanied by consciousness, are equally bound to maintain that those nervous changes in the dog, which correspond with those which underlie sensation in a man, are also unaccompanied by consciousness. In other words, if there is no ground for believing that a dog thinks, neither is there any for believing that he feels.

As is well known, Descartes boldly faced this dilemma, and maintained that all animals were mere machines and entirely devoid of consciousness. But he did not deny, nor can anyone deny, that in this case they are reasoning machines, capable of performing all those operations which are performed by the nervous system of man when he reasons. For even supposing that in man, and in man only, psychosis is superadded to neurosis—the neurosis which is common to both man and animal gives their reasoning processes a fundamental unity. But Descartes' position is open to very serious objections if the evidence that animals feel is insufficient to prove that they really do so. What is the value of the evidence which leads one to believe that one's fellow-man feels? The only evidence in this argument of analogy is the similarity of his structure and of his actions to one's own. And if that is good enough to prove that one's fellow-man feels, surely it is good enough to prove that an ape feels. For the differences of structure and function between men and apes are utterly insufficient to warrant the assumption that while men have those states of consciousness we call sensations apes have nothing of the kind. Moreover, we have as good evidence that apes are capable of emotion and volition as we have that men other than ourselves are. But if apes possess three out of the four kinds of states of consciousness which we discover in ourselves, what possible reason is there for denying them the fourth? If they are capable of sensation, emotion, and volition, why are they to be denied thought (in the sense of predication)?

No answer has ever been given to these questions. And as the law of continuity is as much opposed, as is the common sense of mankind, to the notion that all animals are unconscious machines. it may safely be assumed that no sufficient answer ever will be given to them.

There is every reason to believe that consciousness is a function of nervous matter, when that nervous matter has attained a certain degree of organisation, just as we know the other "actions to which the nervous system ministers," such as reflex action and the like, to be. As I have ventured to state my view of the matter elsewhere, "our thoughts are the expression of molecular changes in that matter of life which is the source of our other vital phenomena."

Mr. Wallace objects to this statement in the following terms :—

"Not having been able to find any clue in Professor Huxley's writings to the steps by which he passes from those vital phenomena, which consist only, in their last analysis, of movements by particles of matter, to those other phenomena which we term thought, sensation, or consciousness ; but, knowing that so positive an expression of opinion from him will have great weight with many persons, I shall endeavour to show, with as much brevity as is compatible with clearness, that this theory is not only incapable of proof, but is also, as it appears to me, inconsistent with accurate conceptions of molecular physics."

With all respect for Mr. Wallace, it appears to me that his remarks are entirely beside the question. I really know nothing whatever, and never hope to know anything, of the steps by which the passage from molecular movement to states of consciousness is effected , and I entirely agree with the sense of the passage which he quotes from Professor Tyndall, apparently imagining that it is in opposition to the view I hold.

All that I have to say is, that, in my belief, consciousness and molecular action are capable of examples that may be required, prove that one form of consciousness, at any rate, is, in the strictest sense, the expression of molecular change, it really is not worth while to pursue the inquiry, whether a fact so easily established is consistent with any particular system of molecular physics or not.

Mr. Wallace, in fact, appears to me to have mixed up two very distinct propositions : the one, the indisputable truth that consciousness is correlated with molecular changes in the organ of consciousness; the other, that the nature of that correlation is known, or can be conceived, which is quite another matter. Mr. Wallace, presumably, believes in that correlation of phenomena which we call cause and effect as firmly as I do. But if he has ever been able to form the faintest notion how a cause gives rise to its effect, all I can say is that I envy him. Take the simplest case imaginable—suppose a ball in motion to impinge upon another ball at rest. I know very well, as a matter of fact, that the ball in motion will communicate some of its motion to the ball at rest, and that the motion of the two balls, after collision, is precisely correlated with the masses of both balls and the amount of motion of the first. But how does this come about ? In what manner can we conceive that the *vis viva* of the first ball passes into the second ? I confess I can no more form any conception of what happens in this case, than I can of what takes place when the motion of

particles of my nervous matter, caused by the
impact of a similar ball gives rise to the state of
consciousness I call pain. In ultimate analysis
everything is incomprehensible, and the whole
object of science is simply to reduce the fundamental incomprehensibilities to the smallest possible number.

But to return to the Quarterly Reviewer. He
admits that animals have "mental images of
sensible objects, combined in all degrees of complexity, as governed by the laws of association."
Presumably, by this confused and imperfect statement the Reviewer means to admit more than the
words imply. For mental images of sensible
objects, even though "combined in all degrees of
complexity," are, and can be, nothing more than
mental images of sensible objects. But judgments, emotions, and volitions cannot by any
possibility be included under the head of "mental
images of sensible objects." If the greyhound
had no better mental endowment than the
Reviewer allows him, he might have the "mental
image" of the "sensible object"—the hare—and
that might be combined with the mental images
of other sensible objects, to any degree of complexity, but he would have no power of judging
it to be at a certain distance from him ; no power
of perceiving its similarity to his memory of a
hare ; and no desire to get at it. Consequently
he would stand stock still, and the noble art of

coursing would have no existence. On the other hand, as that art is largely practised, it follows that greyhounds alone possess a number of mental powers, the existence of which, in any animal, is absolutely denied by the Quarterly Reviewer.

Finally, what are the mental powers which he reserves as the especial prerogative of man? They are two. First, the recognition of "ourselves by ourselves as affected and perceiving.—Self-consciousness."

Secondly. " The reflection upon our sensations and perceptions, and asking what they are and why they are.—Reason."

To the faculty defined in the last sentence, the Reviewer, without assigning the least ground for thus departing from both common usage and technical propriety, applies the name of reason. But if man is not to be considered a reasoning being, unless he asks what his sensations and perceptions are, and why they are, what is a Hottentot, or an Australian " black-fellow "; or what the "swinked hedger " of an ordinary agricultural district? Nay, what becomes of an average country squire or parson? How many of these worthy persons who, as their wont is, read the *Quarterly Review*, would do other than stand agape. if you asked them whether they had ever reflected what their sensations and perceptions are and why they are?

So that if the Reviewer's new definition of reathe merit of that virtue which is unconscious; nay, it is, to my understanding, extremely hard to reconcile Mr. Mivart's dictum with that noble summary of the whole duty of man—" Thou shalt love the Lord thy God with all thy heart, and with all thy soul, and with all thy strength : and thou shalt love thy neighbour as thyself." According to Mr. Mivart's definition, the man who loves God and his neighbour, and, out of sheer love and affection for both, does all he can to please them, is, nevertheless, destitute of a particle of real goodness.

And it further happens that Mr. Darwin, who is charged by Mr. Mivart with being ignorant of the distinction between material and formal goodness, discusses the very question at issue in a passage which is well worth reading (vol i. p. 87), and also comes to a conclusion opposed to Mr. Mivart's axiom. A proposition which has been so much disputed and repudiated, should, under no circumstances, have been thus confidently assumed to be true. For myself, I utterly reject it, inasmuch as the logical consequence of the adoption of any such principle is the denial of all moral value to sympathy and affection. According to Mr. Mivart's axiom, the man who, seeing another struggling in the water, leaps in at the risk of his own life to save him, does that which is " destitute of the most incipient degree of real goodness," unless, as he strips off his coat, he says to himself, " Now, mind, I am going to do this because it is my duty and for no other reason;" and the most beautiful character to which humanity can attain, that of the man who does good without thinking about it, because he loves justice and mercy and is repelled by evil, has no claim on our moral approbation. The denial that a man acts morally because he does not think whether he does so or not, may be put upon the same footing as the denial of the title of an arithmetician to the calculating boy, because he did not know how he worked his sums. If mankind ever generally accept and act upon Mr. Mivart's axiom, they will simply become a set of most unendurable prigs ; but they never have accepted it, and I venture to hope that evolution has nothing so terrible in store for the human race.

But if an action, the motive of which is nothing but affection or sympathy, may be deserving of moral approbation and really good, who that has ever had a dog of his own will deny that animals are capable of such actions ? Mr. Mivart indeed says :—" It may be safely affirmed, however, that there is no trace in brutes of any actions simulating morality which are not explicable by the fear of punishment, by the hope of pleasure, or by personal affection " (p. 221). But it may be affirmed, with equal truth, that there is no trace in men of any actions which are not traceable to the same motives. If a man does anything, he does it either because he fears to be punished if he does not do it, or because he hopes to obtain pleasure and eternal law of human nature that "ginger is hot in the mouth," the assertion has as much foundation of truth as the other, though I think it would be expressed in needlessly pompous language. I must confess that I have never been able to understand why there should be such a bitter quarrel between the intuitionists and the utilitarians. The intuitionist is, after all, only a utilitarian who believes that a particular class of pleasures and pains has an especial importance, by reason of its foundation in the nature of man, and its inseparable connection with his very existence as a thinking being. And as regards the motive of personal affection: Love, as Spinoza profoundly says, is the association of pleasure with that which is loved.[1] Or, to put it to the common sense of mankind, is the gratification of affection a pleasure or a pain? Surely a pleasure. So that whether the motive which leads us to perform an action is the love of our neighbour, or the love of God, it is undeniable that pleasure enters into that motive.

Thus much in reply to Mr. Mivart's arguments. I cannot but think that it is to be regretted that he ekes them out by ascribing to the doctrines of the philosophers with whom he does not agree, logical consequences which have been over and over again proved not to flow from them: and when reason fails him, tries the effect of an injurious

---

[1] "Nempe, Amor nihil aliud est, quam Lætitia, concomitante idea causæ externæ."—*Ethices*, III xiii

nickname. According to the views of Mr. Spencer, Mr. Mill, and Mr. Darwin, Mr. Mivart tells us, " *virtue is a mere kind of retrieving :* " and, that we may not miss the point of the joke, he puts it in italics. But what if it is ? Does that make it less virtue ? Suppose I say that sculpture is a " mere way " of stone-cutting, and painting a " mere way " of daubing canvas, and music a " mere way " of making a noise, the statements are quite true ; but they only show that I see no other method of depreciating some of the noblest aspects of humanity than that of using language in an inadequate and misleading sense about them. And the peculiar inappropriateness of this particular nickname to the views in question, arises from the circumstance which Mr. Mivart would doubtless have recollected, if his wish to ridicule had not for the moment obscured his judgment—that whether the law of evolution applies to man or not, that of hereditary transmission certainly does. Mr. Mivart will hardly deny that a man owes a large share of the moral tendencies which he exhibits to his ancestors ; and the man who inherits a desire to steal from a kleptomaniac, or a tendency to benevolence from a Howard, is, so far as he illustrates hereditary transmission, comparable to the dog who inherits the desire to fetch a duck out of the water from his retrieving sire. So that, evolution, or no evolution, moral qualities are comparable to a "kind of retrieving;" though the comparison, if meant for the purposes of casting obloquy on evolution, does not say much for the fairness of those who make it.

The Quarterly Reviewer and Mr. Mivart base their objections to the evolution of the mental faculties of man from those of some lower animal form upon what they maintain to be a difference in kind between the mental and moral faculties of men and brutes ; and I have endeavoured to show, by exposing the utter unsoundness of their philosophical basis, that these objections are devoid of importance.

The objections which Mr. Wallace brings forward to the doctrine of the evolution of the mental faculties of man from those of brutes by natural causes, are of a different order, and require separate consideration.

If I understand him rightly, he by no means doubts that both the bodily and the mental faculties of man have been evolved from those of some lower animal ; but he is of opinion that some agency beyond that which has been concerned in the evolution of ordinary animals has been operative in the case of man. "A superior intelligence has guided the development of man in a definite direction and for a special purpose, just as man guides the development of many animal and vegetable forms." [1] I understand this

---

[1] "The Limits of Natural Selection as applied to Man" (*loc. cit.* p. 359).

to mean that, just as the rock-pigeon has been
produced by natural causes, while the evolution of
the tumbler from the blue rock has required the
special intervention of the intelligence of man, so
some anthropoid form may have been evolved by
variation and natural selection; but it could never
have given rise to man, unless some superior intelligence had played the part of the pigeon-fancier.

According to Mr. Wallace, "whether we compare the savage with the higher developments of
man, or with the brutes around him, we are alike
driven to the conclusion, that, in his large and
well-developed brain, he possesses an organ quite
disproportioned to his requirements" (p. 343);
and he asks, "What is there in the life of the
savage but the satisfying of the cravings of ap-
petite in the simplest and easiest way? What
thoughts, idea, or actions are there that raise him
many grades above the elephant or the ape?"
(p. 342.) I answer Mr. Wallace by citing a remarkable passage which occurs in his instructive
paper on "Instinct in Man and Animals."

"Savages make long journeys in many directions, and, their
whole faculties being directed to the subject, they gain a wide
and accurate knowledge of the topography, not only of their
own district, but of all the regions round about. Every one
who has travelled in a new direction communicates his knowledge to those who have travelled less, and descriptions of routes
and localities, and minute incidents of travel, form one of the
main staples of conversation around the evening fire. Every
wanderer or captive from another tribe adds to the store of

information, and, as the very existence of individuals and of
whole families and tribes depends upon the completeness of this
knowledge, all the acute perceptive faculties of the adult savage
are directed to acquiring and perfecting it. The good hunter or
warrior thus comes to know the bearing of every hill and mountain range, the directions and junctions of all the streams, the
situation of each tract characterised by peculiar vegetation, not
only within the area he has himself traversed, but perhaps for
a hundred miles around it. His acute observation enables him
to detect the slightest undulations of the surface, the various
changes of subsoil and alterations in the character of the vegetation that would be quite imperceptible to a stranger. His eye is
always open to the direction in which he is going ; the mossy
side of trees, the presence of certain plants under the shade of
rocks, the morning and evening flight of birds, are to him
indications of direction almost as sure as the sun in the heavens "
(pp. 207, 208).

I have seen enough of savages to be able to
declare that nothing can be more admirable than
this description of what a savage has to learn.
But it is incomplete. Add to all this the knowledge which a savage is obliged to gain of the
properties of plants, of the characters and habits
of animals, and of the minute indications by which
their course is discoverable : consider that even an
Australian can make excellent baskets and nets,
and neatly fitted and beautifully balanced spears ,
that he learns to use these so as to be able to
transfix a quartern loaf at sixty yards; and that
very often, as in the case of the American Indians,
the language of a savage exhibits complexities
which a well-trained European finds it difficult to
master : consider that every time a savage tracks

his game he employs a minuteness of observation, and an accuracy of inductive and deductive reasoning which, applied to other matters, would assure some reputation to a man of science, and I think we need ask no further why he possesses such a fair supply of brains.   In complexity and difficulty, I should say that the intellectual labour of a " good hunter or warrior " considerably exceeds that of an ordinary Englishman.   The Civil Service Examiners are held in great terror by young Englishmen ; but even their ferocity never tempted them to require a candidate to possess such a knowledge of a parish as Mr. Wallace justly points out savages may possess of an area a hundred miles or more in diameter.

But suppose, for the sake of argument, that a savage has more brains than seems proportioned to his wants, all that can be said is that the objection to natural selection, if it be one, applies quite as strongly to the lower animals.   The brain of a porpoise is quite wonderful for its mass, and for the development of the cerebral convolutions.   And yet since we have ceased to credit the story of Arion, it is hard to believe that porpoises are much troubled with intellect : and still more difficult is it to imagine that their big brains are only a preparation for the advent of some accomplished cetacean of the future   Surely, again, a wolf must have too much brains, or else how is it that a dog with only the same quantity and form of brain is

able to develop such singular intelligence? The
wolf stands to the dog in the same relation as the
savage to the man; and, therefore, if Mr. Wallace's
doctrine holds good, a higher power must have
superintended the breeding up of wolves from
some inferior stock, in order to prepare them to
become dogs.

Mr. Wallace further maintains that the origin
of some of man's mental faculties by the preservation of useful variations is not possible. Such,
for example, are "the capacity to form ideal conceptions of space and time, of eternity and infinity; the capacity for intense artistic feelings of
pleasure in form, colour, and composition; and for
those abstract notions of form and number which
render geometry and arithmetic possible." "How,"
he asks, "were all or any of these faculties first
developed, when they could have been of no possible use to man in his early stages of barbarism?"

Surely the answer is not far to seek. The
lowest savages are as devoid of any such conceptions as the brutes themselves. What sort of
conceptions of space and time, of form and number, can be possessed by a savage who has not got
so far as to be able to count beyond five or six, who
does not know how to draw a triangle or a circle,
and has not the remotest notion of separating the
particular quality we call form, from the other
qualities of bodies? None of these capacities are
exhibited by men, unless they form part of a

tolerably advanced society. And, in such a society, there are abundant conditions by which a selective influence is exerted in favour of those persons who exhibit an approximation towards the possession of these capacities.

The savage who can amuse his fellows by telling a good story over the nightly fire, is held by them in esteem and rewarded, in one way or another, for so doing—in other words, it is an advantage to him to possess this power. He who can carve a paddle, or the figure-head of a canoe better, similarly profits beyond his duller neighbour. He who counts a little better than others, gets most yams when barter is going on, and forms the shrewdest estimate of the numbers of an opposing tribe. The experience of daily life shows that the conditions of our present social existence exercise the most extraordinarily powerful selective influence in favour of novelists, artists, and strong intellects of all kinds ; and it seems unquestionable that all forms of social existence must have had the same tendency, if we consider the indisputable facts that even animals possess the power of distinguishing form and number, and that they are capable of deriving pleasure from particular forms and sounds. If we admit, as Mr. Wallace does, that the lowest savages are not raised " many grades above the elephant and the ape ; " and if we further admit, as I contend must be admitted, that the conditions of social life tend, powerfully, to give an advantage to those individuals who vary in the direction of intellectual or æsthetic excellence, what is there to interfere with the belief that these higher faculties, like the rest, owe their development to natural selection?

Finally, with respect to the development of the moral sense out of the simple feelings of pleasure and pain, liking and disliking, with which the lower animals are provided, I can find nothing in Mr. Wallace's reasonings which has not already been met by Mr. Mill, Mr. Spencer, or Mr. Darwin.

I do not propose to follow the Quarterly Reviewer and Mr. Mivart through the long string of objections in matters of detail which they bring against Mr. Darwin's views. Every one who has considered the matter carefully will be able to ferret out as many more "difficulties", but he will also, I believe, fail as completely as they appear to me to have done, in bringing forward any fact which is really contradictory of Mr. Darwin's views. Occasionally, too, their objections and criticisms are based upon errors of their own. As, for example, when Mr. Mivart and the Quarterly Reviewer insist upon the resemblances between the eyes of *Cephalopoda* and *Vertebrata*, quite forgetting that there are striking and altogether fundamental differences between them; or when the Quarterly Reviewer corrects Mr. Darwin for saying that the gibbons, " without having been taught, can walk or run upright with tolerable quickness, though they move awkwardly, and much less securely than man." The Quarterly Reviewer says, " This is a little misleading, inasmuch as it is not stated that this upright progression is effected by placing the enormously long arms behind the head, or holding them out backwards as a balance in progression."

Now, before carping at a small statement like this, the Quarterly Reviewer should have made sure that he was quite right. But he happens to be quite wrong. I suspect he got his notion of the manner in which a gibbon walks from a citation in " Man's Place in Nature." But at that time I had not seen a gibbon walk. Since then I have, and I can testify that nothing can be more precise than Mr. Darwin's statement. The gibbon I saw walked without either putting his arms behind his head or holding them out backwards. All he did was to touch the ground with the outstretched fingers of his long arms now and then, just as one sees a man who carries a stick, but does not need one, touch the ground with it as he walks along.

Again, a large number of the objections brought forward by Mr. Mivart and the Quarterly Reviewer apply to evolution in general, quite as much as to the particular form of that doctrine advocated by Mr. Darwin ; or, to their notions of Mr. Darwin's views and not to what they really are. An excellent example of this class of difficulties is to be found in Mr. Mivart's chapter on " Independent Similarities of Structure." Mr. Mivart says that these cannot be explained by an "absolute and pure Darwinian," but " that an innate power and evolutionary law, aided by the corrective action of natural selection, should have furnished like needs with like aids, is not at all improbable" (p. 82).

I do not exactly know what Mr. Mivart means by an "absolute and pure Darwinian;" indeed Mr. Mivart makes that creature hold so many singular opinions that I doubt if I can ever have seen one alive. But I find nothing in his statement of the view which he imagines to be originated by himself, which is really inconsistent with what I understand to be Mr. Darwin's views.

I apprehend that the foundation of the theory of natural selection is the fact that living bodies tend incessantly to vary. This variation is neither indefinite, nor fortuitous, nor does it take place in all directions, in the strict sense of these words.

Accurately speaking, it is not indefinite, nor does it take place in all directions, because it is limited by the general characters of the type to which the organism exhibiting the variation belongs. A whale does not tend to vary in the direction of producing feathers, nor a bird in the direction of developing whalebone. In popular language there is no harm in saying that the

waves which break upon the sea-shore are indefinite, fortuitous, and break in all directions.   In
scientific language, on the contrary, such a statement would be a gross error, inasmuch as every
particle of foam is the result of perfectly definite
forces, operating according to no less definite laws.
In like manner, every variation of a living form,
however minute, however apparently accidental, is
inconceivable except as the expression of the
operation of molecular forces or " powers " resident
within the organism.   And, as these forces certainly
operate according to definite laws, their general
result is, doubtless, in accordance with some general
law which subsumes them all.   And there appears
to be no objection to call this an "evolutionary
law." But nobody is the wiser for doing so, or has
thereby contributed, in the least degree, to the
advance of the doctrine of evolution, the great
need of which is a theory of variation.

When Mr. Mivart tells us that his " aim has
been to support the doctrine that these species
have been evolved by ordinary *natural laws* (for
the most part unknown), aided by the *subordinate*
action of ' natural selection ' " (pp. 332–3), he seems
to be of opinion that his enterprise has the merit
of novelty.   All I can say is that I have never had
the slightest notion that Mr. Darwin's aim is in
any way different from this. If I affirm that
" species have been evolved by variation [1] (a natural

[1] Including under this head hereditary transmission.

process, the laws of which are for the most part unknown), aided by the subordinate action of natural selection," it seems to me that I enunciate a proposition which constitutes the very pith and marrow of the first edition of the "Origin of Species." And what the evolutionist stands in need of just now, is not an iteration of the fundamental principle of Darwinism, but some light upon the questions, What are the limits of variation ? and, If a variety has arisen, can that variety be perpetuated, or even intensified, when selective conditions are indifferent, or perhaps unfavourable to its existence ? I cannot find that Mr. Darwin has ever been very dogmatic in answering these questions. Formerly, he seems to have inclined to reply to them in the negative, while now his inclination is the other way. Leaving aside those broad questions of theology, philosophy, and ethics, by the discussion of which neither the Quarterly Reviewer nor Mr. Mivart can be said to have damaged Darwinism—whatever else they have injured—this is what their criticisms come to. They confound a struggle for some rifle-pits with an assault on the fortress.

In some respects, finally, I can only characterise the Quarterly Reviewer's treatment of Mr. Darwin as alike unjust and unbecoming. Language of this strength requires justification, and on that ground I add the remarks which follow.

The Quarterly Reviewer opens his essay by a

careful enumeration of all those points upon which,
during the course of thirteen years of incessant
labour, Mr. Darwin has modified his opinions. It
has often and justly been remarked, that what
strikes a candid student of Mr. Darwin's works is
not so much his industry, his knowledge, or even
the surprising fertility of his inventive genius;
but that unswerving truthfulness and honesty
which never permit him to hide a weak place, or
gloss over a difficulty, but lead him, on all occasions, to point out the weak places in his own
armour, and even sometimes, it appears to me, to
make admissions against himself which are quite
unnecessary. A critic who desires to attack Mr.
Darwin has only to read his works with a desire to
observe, not their merits, but their defects, and he
will find, ready to hand, more adverse suggestions
than are likely ever to have suggested themselves
to his own sharpness, without Mr. Darwin's self-
denying aid.

Now this quality of scientific candour is not so
common that it needs to be discouraged; and it
appears to me to deserve other treatment than
that adopted by the Quarterly Reviewer, who deals
with Mr. Darwin as an Old Bailey barrister deals
with a man against whom he wishes to obtain a
conviction, *per fas aut nefas*, and opens his case
by endeavouring to create a prejudice against the
prisoner in the minds of the jury. In his eager-
ness to carry out this laudable design, the Quarterly

Reviewer cannot even state the history of the
doctrine of natural selection without an oblique
and entirely unjustifiable attempt to depreciate
Mr. Darwin. " To Mr. Darwin," says he, " and
(through Mr. Wallace's reticence) to Mr. Darwin
alone, is due the credit of having first brought it
prominently forward and demonstrated its truth."
No one can less desire than I do, to throw a doubt
upon Mr. Wallace's originality, or to question his
claim to the honour of being one of the originators
of the doctrine of natural selection ; but the statement that Mr. Darwin has the sole credit of
originating the doctrine because of Mr. Wallace's
reticence is simply ridiculous. The proof of this
is, in the first place, afforded by Mr. Wallace himself, whose noble freedom from petty jealousy in
this matter smaller folk would do well to imitate,
and who writes thus :—" I have felt all my life,
and I still feel, the most sincere satisfaction
that Mr. Darwin had been at work long before
me and that it was not left for me to attempt to
write the ' Origin of Species.' I have long since
measured my own strength, and know well that it
would be quite unequal to that task." So that if
there was any reticence at all in the matter, it was
Mr. Darwin's reticence during the long twenty
years of study which intervened between the conception and the publication of his theory, which
gave Mr. Wallace the chance of being an independent discoverer of the importance of natural

VI

EVOLUTION IN BIOLOGY

[1878]

IN the former half of the eighteenth century, the term " evolution " was introduced into biological writings, in order to denote the mode in which some of the most eminent physiologists of that time conceived that the generations of living things took place; in opposition to the hypothesis advocated, in the preceding century, by Harvey in that remarkable work [1] which would give him a claim to rank among the founders of biological science, even had he not been the discoverer of the circulation of the blood.

One of Harvey's prime objects is to defend and establish, on the basis of direct observation, the opinion already held by Aristotle ; that, in the higher animals at any rate, the formation of the

---

[1] The *Exercitationes de Generatione Animalium*, which Dr. George Ent extracted from him and published in 1651.

new organism by the process of generation takes
place, not suddenly, by simultaneous accretion of
rudiments of all, or of the most important, of the
organs of the adult; nor by sudden metamorphosis
of a formative substance into a miniature of the
whole, which subsequently grows; but by *epigenesis,*
or successive differentiation of a relatively homo-
geneous rudiment into the parts and structures
which are characteristic of the adult.

"Et primo, quidem, quoniam per *epigenesin* sive partium
superexorientium additamentum pullum fabricari certum est:
quænam pars ante alias omnes exstruatur, et quid de illa ejusque
generandi modo observandum veniat, dispiciemus. Ratum sane
est et in ovo manifestè apparet quod *Aristoteles* de perfectorum
animalium generatione enuntiat: nimirum, non omnes partes
simul fieri, sed ordine aliam post aliam; primumque existere
particulam genitalem, cujus virtute postea (tanquam ex principio
quodam) reliquæ omnes partes prosiliant. Qualem in plantarum
seminibus (fabis, putà, aut glandibus) gemmam sive apicem pro-
tuberantem cernimus, totius futuræ arboris principium. *Estque
hæc particula velut filius emancipatus seorsumque collocatus, et
principium per se vivens; unde postea membrorum ordo describ-
itur; et quæcunque ad absolvendum animal pertinent, dispon-
untur.*[1] Quoniam enim *nulla pars se ipsam generat; sed post-
quam generata est, se ipsam jam auget; ideo eam primùm oriri
necesse est, quæ principium augendi contineat (sive enim planta,
sive animal est, æque omnibus inest quod vim habeat vegetandi,
sive nutriendi),*[2] simulque reliquas omnes partes suo quamque
ordine distinguat et formet; proindeque in eadem primogenita
particula anima primario inest, sensus, motusque, et totius vitæ
auctor et principium." (Exercitatio 51.)

---

[1] *De Generatione Inimalium,* lib. ii. cap. x.
[2] *De Generatione,* lib. ii. cap. iv.

Harvey proceeds to contrast this view with that of the " Medici," or followers of Hippocrates and Galen, who, " badly philosophising," imagined that the brain, the heart, and the liver were simultaneously first generated in the form of vesicles ; and, at the same time, while expressing his agreement with Aristotle in the principle of epigenesis, he maintains that it is the blood which is the primal generative part, and not, as Aristotle thought, the heart.

In the latter part of the seventeenth century, the doctrine of epigenesis, thus advocated by Harvey, was controverted, on the ground of direct observation, by Malpighi, who affirmed that the body of the chick is to be seen in the egg, before the *punctum sanguineum* makes it appearance. But, from this perfectly correct observation a conclusion which is by no means warranted was drawn ; namely, that the chick, as a whole, really exists in the egg antecedently to incubation ; and that what happens in the course of the latter process is no addition of new parts, " alias post alias natas," as Harvey puts it, but a simple expansion, or unfolding, of the organs which already exist, though they are too small and inconspicuous to be discovered. The weight of Malpighi's observations therefore fell into the scale of that doctrine which Harvey terms *metamorphosis*, in contradistinction to epigenesis.

The views of Malphigi were warmly welcomed,

on philosophical grounds, by Leibnitz,[1] who found
in them a support to his hypothesis of monads,
and by Malebranche ;[2] while, in the middle of the
eighteenth century, not only speculative considerations, but a great number of new and interesting
observations on the phenomena of generation, led
the ingenious Bonnet, and Haller,[3] the first physiologist of the age, to adopt, advocate, and extend
them.

[1] "Cependant, pour revenir aux formes ordinaires ou aux
âmes matérielles, cette durée qu'il leur faut attribuer à la place
de celle qu'on avoit attribuée aux atomes pourroit faire douter
si elles ne vont pas de corps en corps ; ce qui seroit la mé-
tempsychose, à peu près comme quelques philosophes ont cru la
transmission du mouvement et celle des espèces. Mais cette
imagination est bien éloignée de la nature des choses.   Il n'y a
point de tel passage ; et c'est ici où les transformations de
Messieurs Swammerdam, Malpighi, et Leewenhoek, qui sont
des plus excellens observateurs de notre tems, sont venues à mon
secours, et m'ont fait admettre plus aisément, que l'animal, et
toute autre substance organisée ne commence point lorsque nous
le croyons, et que sa generation apparente n'est qu'une dé-
veloppement et une espèce d'augmentation. Aussi ai je remarqué
que l'auteur de la *Recherche de la Verité*, M. Regis, M. Hart-
soeker, et d'autres habiles hommes n'ont pas été fort éloignés
de ce sentiment."   Leibnitz, *Système Nouveau de la Nature*,
1695.   The doctrine of "Emboîtement" is contained in the
Considérations sur le Principe de Vie, 1705 ; the preface to the
Theodicée, 1710 ; and the *Principes de la Nature et de la Grace*
(§ 6), 1718.

[2] "Il est vrai que la pensée la plus raisonnable et la plus
conforme à l'experience sur cette question très difficile de la
formation du fœtus ; c'est que les enfans sont déja presque tout
formés avant même l'action par laquelle ils sont conçus ; et que
leurs mères ne font que leur donner l'accroissement ordinaire
dans le temps de la grossesse."   *De la Recherche de la Verité*,
livre ii. chap. vii. p. 334, 7th ed., 1721.

[3] The writer is indebted to Dr. Allen Thomson for reference
to the evidence contained in a note to Haller's edition of Boer-
haave's *Prælectiones Academicæ*, vol. v. pt. ii. p. 497, published
in 1744, that Haller originally advocated epigenesis.

Bonnet affirms that, before fecundation, the hen's egg contains an excessively minute but complete chick ; and that fecundation and incubation simply cause this germ to absorb nutritious matters, which are deposited in the interstices of the elementary structures of which the miniature chick, or germ, is made up. The consequence of this intussusceptive growth is the " development " or " evolution " of the germ into the visible bird. Thus an organised individual (*tout organisé*) " is a composite body consisting of the original, or *elementary*, parts and of the matters which have been associated with them by the aid of nutrition ; " so that, if these matters could be extracted from the individual (*tout*), it would, so to speak, become concentrated in a point, and would thus be restored to its primitive condition of a *germ ;* "just as by extracting from a bone the calcareous substance which is the source of its hardness, it is reduced to its primitive state of gristle or membrane." [1]

"Evolution" and "development" are, for Bonnet, synonymous terms ; and since by " evolution " he means simply the expansion of that which was invisible into visibility, he was naturally led to the conclusion, at which Leibnitz had arrived by a different line of reasoning, that no such thing as generation, in the proper sense of the word, exists in Nature. The growth of an

[1] *Considérations sur les Corps organisés*, chap. x.

organic being is simply a process of enlargement
as a particle of dry gelatine may be swelled
up by the intussusception of water; its death
is a shrinkage, such as the swelled jelly might
undergo on desiccation. Nothing really new is
produced in the living world, but the germs which
develop have existed since the beginning of things;
and nothing really dies, but, when what we call
death takes place, the living thing shrinks back
into its germ state.[1]

The two parts of Bonnet's hypothesis, namely
the doctrine that all living things proceed from
pre-existing germs, and that these contain, one

---

[1] Bonnet had the courage of his opinions, and in the
Palingénésie Philosophique, part vi. chap. iv., he develops a
hypothesis which he terms "évolution naturelle;" and which,
making allowance for his peculiar views of the nature of
generation, bears no small resemblance to what is understood
by "evolution" at the present day :—

"Si la volonté divine a créé par un seul Acte l'Universalité
des êtres, d'où venoient ces plantes et ces animaux dont Moyse
nous decrit la Production au troisieme et au cinquieme jour du
renouvellement de notre monde?

" Abuserois-je de la liberté de conjectures si je disois, que les
Plantes et les Animaux qui existent aujourd'hui sont parvenus
par une sorte d'evolution naturelle des Etres organisés qui
peuplaient ce premier Monde, sorti immédiatement des MAINS
du CREATEUR? . . .

"Ne supposons que trois révolutions. La Terre vient de sortir
des MAINS du CREATEUR. Des causes preparées par sa SAGESSE
font développer de toutes parts les Germes. Les Etres organisés
commencent à jouir de l'existence. Ils étoient probablement
alors bien différens de ce qu'ils sont aujourd'hui. Ils l'etoient
autant que ce premier Monde différoit de celui que nous habitons.
Nous manquons de moyens pour juger de ces dissemblances,
et peut-être que le plus habile Naturaliste qui auroit été placé
dans ce premier Monde y auroit entièrement méconnu nos Plantes
et nos Animaux."

inclosed within the other, the germs of all future living things, which is the hypothesis of "*emboîte-ment ;*" and the doctrine that every germ contains in miniature all the organs of the adult, which is the hypothesis of evolution or development, in the primary senses of these words, must be carefully distinguished. In fact, while holding firmly by the former, Bonnet more or less modified the latter in his later writings, and, at length, he admits that a "germ" need not be an actual miniature of the organism; but that it may be merely an "original preformation" capable of producing the latter.[1]

But, thus defined, the germ is neither more nor less than the "particula genitalis" of Aristotle, or the "primordium vegetale" or "ovum" of Harvey; and the "evolution" of such a germ would not be distinguishable from "epigenesis."

Supported by the great authority of Haller, the doctrine of evolution, or development, prevailed throughout the whole of the eighteenth century, and Cuvier appears to have substantially adopted Bonnet's later views, though probably he would not have gone all lengths in the direction of "emboîtement." In a well-known note to Laurillard's "Éloge," prefixed to the last edition

---

[1] "Ce mot (germe) ne désignera pas seulement un corps organisé *réduit en petit ;* il désignera encore toute espèce de *pré-formation originelle dont un Tout organique peut résulter comme de son principe immédiat.*"—*Palingénésie Philosophique,* part x. chap. ii.

of the " Ossemens fossiles," the " radical de l'être "
is much the same thing as Aristotle's "particula
genitalis " and Harvey's " ovum." [1]

Bonnet's eminent contemporary, Buffon, held
nearly the same views with respect to the nature
of the germ, and expresses them even more confidently.

" Ceux qui ont cru que le cœur étoit le premier formé, se sont
trompés ; ceux qui disent que c'est le sang se trompent aussi :
tout est formé en même temps.   Si l'on ne consulte que l'observation, le poulet se voit dans l'œuf avant qu'il ait été couvé." [2]

" J'ai ouvert une grande quantité d'œufs à differens temps
avant et après l'incubation, et je me suis convaincu par mes
yeux que le poulet existe en entier dans le milieu de la cicatricule
au moment qu'il sort du corps de la poule." [3]

The " moule intérieur " of Buffon is the aggre-
gate of elementary parts which constitute the
individual, and is thus the equivalent of Bonnet's
germ,[4] as defined in the passage cited above.
But Buffon further imagined that innumerable
" molecules organiques " are dispersed throughout
the world, and that alimentation consists in the

---

[1] " M. Cuvier considérant que tous les êtres organisés sont
dérivés de parens, et ne voyant dans la nature aucune force
capable de produire l'organisation, croyait à la pré-existence
des germes ; non pas à la pré-existence d'un être tout formé,
puisqu'il est bien évident que ce n'est que par des développemens
successifs que l'être acquiert sa forme ; mais, si l'on peut
s'exprimer ainsi, à la pré-existence du *radical de l'être*, radical
qui existe avant que la série des évolutions ne commence, et qui
remonte certainement, suivant la belle observation de Bonnet, à
plusieurs generations."—Laurillard, *Éloge de Cuvier*, note 12.

[2] *Histoire Naturelle*, tom. ii. ed. ii. 1750, p. 350.

[3] *Ibid.*, p. 351.   [4] See particularly Buffon, *l.c.* p. 41.

appropriation by the parts of an organism of those
molecules which are analogous to them. Growth,
therefore, was, on this hypothesis, a process
partly of simple evolution, and partly of what has
been termed "syngenesis." Buffon's opinion is,
in fact, a sort of combination of views, essentially
similar to those of Bonnet, with others, somewhat
similar to those of the "Medici" whom Harvey
condemns. The "molecules organiques" are
physical equivalents of Leibnitz's "monads."

It is a striking example of the difficulty of
getting people to use their own powers of investigation accurately, that this form of the doctrine of
evolution should have held its ground so long;
for it was thoroughly and completely exploded,
not long after its enunciation, by Casper Friederich
Wolff, who in his "Theoria Generationis," published in 1759, placed the opposite theory of
epigenesis upon the secure foundation of fact,
from which it has never been displaced. But
Wolff had no immediate successors. The school
of Cuvier was lamentably deficient in embryo-
logists; and it was only in the course of the first
thirty years of the present century, that Prévost
and Dumas in France, and, later on, Döllinger,
Pander, Von Bär, Rathke, and Remak in Germany,
founded modern embryology; while, at the same
time, they proved the utter incompatibility of the
hypothesis of evolution, as formulated by Bonnet
and Haller, with easily demonstrable facts.

Nevertheless, though the conceptions originally denoted by "evolution" and "development" were shown to be untenable, the words retained their application to the process by which the embryos of living beings gradually make their appearance; and the terms "Development," "Entwickelung," and "Evolutio," are now indiscriminately used for the series of genetic changes exhibited by living beings, by writers who would emphatically deny that "Development" or "Entwickelung" or "Evolutio," in the sense in which these words were usually employed by Bonnet or by Haller, ever occurs.

Evolution, or development, is, in fact, at present employed in biology as a general name for the history of the steps by which any living being has acquired the morphological and the physiological characters which distinguish it. As civil history may be divided into biography, which is the history of individuals, and universal history, which is the history of the human race, so evolution falls naturally into two categories—the evolution of the individual, and the evolution of the sum of living beings. It will be convenient to deal with the modern doctrine of evolution under these two heads.

I. *The Evolution of the Individual.*

No exception is at this time, known to the general law, established upon an immense multitude of direct observations, that every living thing

precursor and model, with the generous respect with
which one genuine worker should regard another
—that such germs may arise by a process of
" equivocal generation " out of not-living matter;
and the aphorism so commonly ascribed to him,
"*omne vivum ex ovo*," and which is indeed a fair
summary of his reiterated assertions, though
incessantly employed against the modern advocates of spontaneous generation, can be honestly
so used only by those who have never read a
score of pages of the " Exercitationes." Harvey,
in fact, believed as implicitly as Aristotle did in the
equivocal generation of the lower animals. But,
while the course of modern investigation has only
brought out into greater prominence the accuracy
of Harvey's conception of the nature and mode of
development of germs, it has as distinctly tended
to disprove the occurrence of equivocal generation,
or abiogenesis, in the present course of nature.
In the immense majority of both plants and
animals, it is certain that the germ is not merely
a body in which life is dormant or potential, but
that it is itself simply a detached portion of
the substance of a pre-existing living body ; and
the evidence has yet to be adduced which will
satisfy any cautious reasoner that " omne vivum
ex vivo " is not as well-established a law of
the existing course of nature as " omne vivum
ex ovo."

In all instances which have yet been investigated, the substance of this germ has a peculiar chemical composition, consisting of at fewest four elementary bodies, viz., carbon, hydrogen, oxygen, and nitrogen, united into the ill-defined compound known as protein, and associated with much water, and very generally, if not always, with sulphur and phosphorus in minute proportions. Moreover, up to the present time, protein is known only as a product and constituent of living matter. Again, a true germ is either devoid of any structure discernible by optical means, or, at most, it is a simple nucleated cell.[1]

In all cases the process of evolution consists in a succession of changes of the form, structure, and functions of the germ, by which it passes, step by step, from an extreme simplicity, or relative homogeneity, of visible structure, to a greater or less degree of complexity or heterogeneity; and the course of progressive differentiation is usually accompanied by growth, which is effected by intussusception. This intussusception, however, is a very different process from that imagined either by Buffon or by Bonnet. The substance by the addition of which the germ is enlarged is in no case simply absorbed, ready-made, from the not-living world and packed between the elementary constituents of the germ, as Bonnet imagined;

---

[1] In some cases of sexless multiplication the germ is a cell-aggregate—if we call germ only that which is already detached from the parent organism.

still less does it consist of the "molecules organiques" of Buffon. The new material is, in great
measure, not only absorbed but assimilated, so
that it becomes part and parcel of the molecular
structure of the living body into which it enters.
And, so far from the fully developed organism
being simply the germ *plus* the nutriment which
it has absorbed, it is probable that the adult contains neither in form, nor in substance, more than
an inappreciable fraction of the constituents of
the germ, and that it is almost, if not wholly,
made up of assimilated and metamorphosed
nutriment.   In the great majority of cases, at
any rate, the full-grown organism becomes what
it is by the absorption of not-living matter, and
its conversion into living matter of a specific type.
As Harvey says (Ex. 45), all parts of the body
are nourished "ab eodem succo alibili, aliter
aliterque cambiato," "ut plantæ omnes ex eodem
communi nutrimento (sive rore seu terræ
humore)."

In all animals and plants above the lowest the
germ is a nucleated cell, using that term in its
broadest sense ; and the first step in the process
of the evolution of the individual is the division
of this cell into two or more portions.   The process of division is repeated, until the organism,
from being unicellular, becomes multicellular.
The single cell becomes a cell-aggregate ; and it
is to the growth and metamorphosis of the cells

of the cell-aggregate thus produced, that all the
organs and tissues of the adult owe their origin.

In certain animals belonging to every one of
the chief groups into which the *Metazoa* are
divisible, the cells of the cell-aggregate which
results from the process of yelk-division, and
which is termed a *morula*, diverge from one
another in such a manner as to give rise to a
central space, around which they dispose themselves as a coat or envelope; and thus the morula
becomes a vesicle filled with fluid, the *planula*.
The wall of the planula is next pushed in on one
side, or invaginated, whereby it is converted into
a double-walled sac with an opening, the *blasto-
pore*, which leads into the cavity lined by the
inner wall. This cavity is the primitive alimentary cavity or *archenteron;* the inner or invaginated layer is the *hypoblast;* the outer the
epiblast; and the embryo, in this stage, is termed
a *gastrula*. In all the higher animals a layer of
cells makes its appearance between the hypoblast
and the epiblast, and is termed the *mesoblast*. In
the further course of development the epiblast
becomes the ectoderm or epidermic layer of the
body; the hypoblast becomes the epithelium of
the middle portion of the alimentary canal; and
the mesoblast gives rise to all the other tissues,
except the central nervous system, which originates from an ingrowth of the epiblast.

With more or less modification in detail, the

embryo has been observed to pass through these successive evolutional stages in sundry Sponges, Cœlenterates, Worms, Echinoderms, Tunicates, Arthropods, Mollusks, and Vertebrates; and there are valid reasons for the belief that all animals of higher organisation than the *Protozoa* agree in the general character of the early stages of their individual evolution. Each, starting from the condition of a simple nucleated cell, becomes a cell-aggregate; and this passes through a condition which represents the gastrula stage, before taking on the features distinctive of the group to which it belongs. Stated in this form, the "gastræa theory" of Haeckel appears to the present writer to be one of most important and best founded of recent generalisations. So far as individual plants and animals are concerned, therefore, evolution is not a speculation but a fact; and it takes place by epigenesis.

"Animal . . . per *epigenesin* procreatur, materiam simul attrahit, parat, concoquit, et eâdem utitur ; formatur simul et augetur . . . primum futuri corporis concrementum . . . prout augetur, dividitur sensim et distinguitur in partes, non simul omnes, sed alias post alias natas, et ordine quasque suo emergentes."[1]

In these words, by the divination of genius, Harvey, in the seventeenth century, summed up the outcome of the work of all those who, with appliances he could not dream of, are continuing his labours in the nineteenth century.

[1] Harvey, *Exercitationes de Generatione.* Ex. 45, "Quænam sit pulli materia et quomodo fiat in Ovo."

Nevertheless, though the doctrine of epigenesis, as understood by Harvey, has definitively triumphed over the doctrine of evolution, as understood by his opponents of the eighteenth century, it is not impossible that, when the analysis of the process of development is carried still further, and the origin of the molecular components of the physically gross, though sensibly minute, bodies which we term germs is traced, the theory of development will approach more nearly to metamorphosis than to epigenesis. Harvey thought that impregnation influenced the female organism as a contagion ; and that the blood, which he conceived to be the first rudiment of the germ, arose in the clear fluid of the " colliquamentum " of the ovum by a process of concrescence, as a sort of living precipitate. We now know, on the contrary, that the female germ or ovum, in all the higher animals and plants, is a body which possesses the structure of a nucleated cell; that impregnation consists in the fusion of the substance [1] of another more or less modified nucleated cell, the male germ, with the ovum ; and that the structural components of the body of the embryo are all derived, by a process of division, from the coalesced male and female germs. Hence it is conceivable, and indeed probable, that every part of the adult contains molecules, derived both from the male and

[1] [At any rate of the nuclei of the two germ-cells. 1893].

from the female parent ; and that, regarded as a
mass of molecules, the entire organism may be compared to a web of which the warp is derived from the
female and the woof from the male. And each of
these may constitute one individuality, in the same
sense as the whole organism is one individual, although the matter of the organism has been constantly changing. The primitive male and female
molecules may play the part of Buffon's "moules
organiques," and mould the assimilated nutriment,
each according to its own type, into innumerable
new molecules. From this point of view the process,
which, in its superficial aspect, is epigenesis, appears
in essence, to be evolution, in the modified sense
adopted in 'Bonnet's later writings; and development is merely the expansion of a potential organism or " original preformation " according to fixed
laws.

II. *The Evolution of the Sum of Living Beings.*

The notion that all the kinds of animals and
plants may have come into existence by the growth
and modification of primordial germs is as old as
speculative thought; but the modern scientific
form of the doctrine can be traced historically to
the influence of several converging lines of philosophical speculation and of physical observation,
none of which go farther back than the seventeenth century. These are :—

1. The enunciation by Descartes of the conception that the physical universe, whether living or not living, is a mechanism, and that, as such, it is explicable on physical principles.

2. The observation of the gradations of structure, from extreme simplicity to very great complexity, presented by living things, and of the relation of these graduated forms to one another.

3. The observation of the existence of an analogy between the series of gradations presented by the species which compose any great group of animals or plants, and the series of embryonic conditions of the highest members of that group.

4. The observation that large groups of species of widely different habits present the same fundamental plan of structure; and that parts of the same animal or plant, the functions of which are very different, likewise exhibit modifications of a common plan.

5. The observation of the existence of structures, in a rudimentary and apparently useless condition, in one species of a group, which are fully developed and have definite functions in other species of the same group.

6. The observation of the effects of varying conditions in modifying living organisms.

7. The observation of the facts of geographical distribution.

8. The observation of the facts of the geological succession of the forms of life.

1. Notwithstanding the elaborate disguise which fear of the powers that were led Descartes to throw over his real opinions, it is impossible to read the "Principes de la Philosophie" without acquiring the conviction that this great philosopher held that the physical world and all things in it, whether living or not living, have originated by a process of evolution, due to the continuous operation of purely physical causes, out of a primitive relatively formless matter.[1]

The following passage is especially instructive :—

"Et tant s'en faut que je veuille que l'on croie toutes les choses que j'écrirai, que même je pretends en proposer ici quelques unes que je crois absolument être fausses ; à savoir, je ne doute point que le monde n'ait été créé au commencement avec autant de perfection qu'il en a ; en sorte que le soleil, la terre, la lune, et les étoiles ont été dès lors ; et que la terre n'a pas eu seulement en soi les semences des plantes, mais que les plantes même en ont couvert une partie ; et qu' Adam et Eve n'ont pas été créés enfans mais en âge d'hommes parfaits. La religion chrétienne veut que nous le croyons ainsi, et la raison naturelle nous persuade entièrement cette vérité ; car si nous considérons la toute puissance de Dieu, nous devons juger que tout ce qu'il a fait a eu dès le commencement toute la perfection qu'il devoit avoir. Mais néanmoins, comme on connôitroit beaucoup mieux quelle a été la nature d'Adam et celle des arbres de Paradis si on avoit examiné comment les enfants se forment peu à peu dans le ventre de leurs mères et comment les plantes sortent de leurs semences, que si on avoit seulement considéré quels ils ont été quand Dieu les a créés : tout de même, nous ferons mieux entendre quelle est

---

[1] As Buffon has well said :—" L'idée de ramener l'explication de tous les phénomènes à des principes mecaniques est assurement grande et belle, ce pas est le plus hardi qu'on peut faire en philosophie, et c'est Descartes qui l'a fait."—*l.c.* p. 50.

généralement la nature de toutes les choses qui sont au monde si
nous pouvons imaginer quelques principes qui soient fort intelli-
gibles et fort simples, desquels nous puissions voir clairement que
les astres et la terre et enfin tout ce monde visible auroit pu être
produit ainsi que de quelques semences (bien que nous sachions
qu'il n'a pas été produit en cette façon) que si nous la decrivions
seulement comme il est, ou bien comme nous croyons qu'il a été
créé.   Et parceque je pense avoir trouvé des principes qui sont
tels, je tacherai ici de les expliquer." [1]

If we read between the lines of this singular
exhibition of force of one kind and weakness of
another, it is clear that Descartes believed that he
had divined the mode in which the physical universe had been evolved ; and the "Traité de
l'Homme," and the essay " Sur les Passions " afford
abundant additional evidence that he sought for,
and thought he had found, an explanation of the
phenomena of physical life by deduction from
purely physical laws.

Spinoza abounds in the same sense, and is as
usual perfectly candid—

" Naturæ leges et regulæ, secundum quas omnia fiunt et ex
unis formis in alias mutantur, sunt ubique et semper eadem." [2]

Leibnitz's doctrine of continuity necessarily led
him in the same direction ; and, of the infinite
multitude of monads with which he peopled the
world, each is supposed to be the focus of an endless process of evolution and involution.   In the

[1] *Principes de la Philosophie*, Troisième partie, § 45.
[2] *Ethices*, Pars tertia, Præfatio.

"Protogæa," xxvi., Leibnitz distinctly suggests the mutability of species—

"Alii mirantur in saxis passim species videri quas vel in orbe cognito, vel saltem in vicinis locis frustra quæras. 'Ita Cornua Ammonis,' quæ ex nautilorum numero habeantur, passim et forma et magnitudine (nam et pedali diametro aliquando reperiuntur) ab omnibus illis naturis discrepare dicunt, quas præbet mare. Sed quis absconditos ejus recessus aut subterraneas abyssos pervestigavit? quam multa nobis animalia antea ignota offert novus orbis? Et credibile est per magnas illas conversiones etiam animalium species plurimum immutatas."

Thus, in the end of the seventeenth century, the seed was sown which has, at intervals, brought forth recurrent crops of evolutional hypotheses, based, more or less completely, on general reasonings.

Among the earliest of these speculations is that put forward by Benoit de Maillet in his "Telliamed," which, though printed in 1735, was not published until twenty-three years later. Considering that this book was written before the time of Haller, or Bonnet, or Linnæus, or Hutton, it surely deserves more respectful consideration than it usually receives. For De Maillet not only has a definite conception of the plasticity of living things, and of the production of existing species by the modification of their predecessors; but he clearly apprehends the cardinal maxim of modern geological science, that the explanation of the structure of the globe is to be sought in the deductive application to geological phenomena of the principles established inductively by the study of the present course of nature. Somewhat later, Maupertuis[1] suggested a curious hypothesis as to the causes of variation, which he thinks may be sufficient to account for the origin of all animals from a single pair. Robinet [2] followed out much the same line of thought as De Maillet, but less soberly ; and Bonnet's speculations in the " Palingénésie," which appeared in 1769, have already been mentioned. Buffon (1753-1778), at first a partisan of the absolute immutability of species, subsequently appears to have believed that larger or smaller groups of species have been produced by the modification of a primitive stock ; but he contributed nothing to the general doctrine of evolution.

Erasmus Darwin (" Zoonomia," 1794), though a zealous evolutionist, can hardly be said to have made any real advance on his predecessors ; and, notwithstanding that Goethe (1791-4) had the advantage of a wide knowledge of morphological facts, and a true insight into their signification, while he threw all the power of a great poet into the expression of his conceptions, it may be questioned whether he supplied the doctrine of evolu-

[1] *Système de la Nature.* " Essai sur la Formation des Corps Organisés," 1751, xiv.
[2] *Considérations Philosophiques sur la gradation naturelle des formes de l'être ; ou les essais de la nature qui apprend à faire l'homme,* 1768.

tion with a firmer scientific basis than it already possessed. Moreover, whatever the value of Goethe's labours in that field, they were not published before 1820, long after evolutionism had taken a new departure from the works of Treviranus and Lamarck—the first of its advocates who were equipped for their task with the needful large and accurate knowledge of the phenomena of life, as a whole. It is remarkable that each of these writers seems to have been led, independently and contemporaneously, to invent the same name of " Biology " for the science of the phenomena of life ; and thus, following Buffon, to have recognised the essential unity of these phenomena, and their contradistinction from those of inanimate nature. And it is hard to say whether Lamarck or Treviranus has the priority in propounding the main thesis of the doctrine of evolution ; for though the first volume of Treviranus's " Biologie " appeared only in 1802, he says, in the preface to his later work, the " Erscheinungen und Gesetze des organischen Lebens," dated 1831, that he wrote the first volume of the " Biologie " " nearly five-and-thirty years ago," or about 1796.

Now, in 1794, there is evidence that Lamarck held doctrines which present a striking contrast to those which are to be found in the " Philosophie Zoologique," as the following passages show :—

" 685. Quoique mon unique objet dans cet article n'ait été que de traiter de la cause physique de l'entretien de la vie des êtres

organiques, malgré cela j'ai osé avancer en débutant, que l'existence de ces êtres étonnants n'appartiennent nullement à la
nature ; que tout ce qu'on peut entendre par le mot *nature*, ne
pouvoit donner la vie, c'est-à-dire, que toutes les qualités de la
matière, jointes à toutes les circonstances possibles, et même à
l'activité répandue dans l'univers, ne pouvaient point produire
un être muni du mouvement organique, capable de reproduire
son semblable, et sujet à la mort.

"686. Tous les individus de cette nature, qui existent, pro-
viennent d'individus semblables qui tous ensemble constituent
l'espèce entière. Or, je crois qu'il est aussi impossible à l'homme
de connôitre la cause physique du premier individu de chaque
espèce, que d'assigner aussi physiquement la cause de l'existence
de la matière ou de l'univers entier. C'est au moins ce que le
résultat de mes connaissances et de mes réflexions me portent à
penser. S'il existe beaucoup de variétés produites par l'effet des
circonstances, ces variétés ne dénaturent point les espèces ; mais
on se trompe, sans doute souvent, en indiquant comme espèce, ce
qui n'est que variété ; et alors je sens que cette erreur peut tirer
à conséquence dans les raisonnements que l'on fait sur cette
matière." [1]

The first three volumes of Treviranus's "Biologie," which contain his general views of
evolution, appeared between 1802 and 1805. The
"Recherches sur l' organisation des corps vivants,"
in which the outlines of Lamarck's doctrines are
given, was published in 1802 ; but the full develop-

[1] *Recherches sur les causes des principaux faits physiques*,
par J. B. Lamarck. Paris. Seconde année de la République.
In the preface, Lamarck says that the work was written in 1776,
and presented to the Academy in 1780 ; but it was not published
before 1794, and, at that time, it presumably expressed Lamarck's
mature views. It would be interesting to know what brought
about the change of opinion manifested in the *Recherches sur
l'organisation des corps vivants*, published only seven years
later.

ment of his views, in the "Philosophie Zoologique," did not take place until 1809.

The "Biologie" and the "Philosophie Zoologique" are both very remarkable productions, and are still worthy of attentive study, but they fell upon evil times. The vast authority of Cuvier was employed in support of the traditionally respectable hypotheses of special creation and of catastrophism; and the wild speculations of the "Discours sur les Révolutions de la Surface du Globe" were held to be models of sound scientific thinking, while the really much more sober and philosophical hypotheses of the "Hydrogeologie" were scouted. For many years it was the fashion to speak of Lamarck with ridicule, while Treviranus was altogether ignored.

Nevertheless, the work had been done. The conception of evolution was henceforward irrepressible, and it incessantly reappears, in one shape or another,[1] up to the year 1858, when Mr. Darwin and Mr. Wallace published their "Theory of Natural Selection." The "Origin of Species" appeared in 1859; and it is within the knowledge of all whose memories go back to that time, that, henceforward, the doctrine of evolution has assumed a position and acquired an importance which it never before possessed. In the "Origin of Species," and in his other numerous and

[1] See the "Historical Sketch" prefixed to the last edition of the *Origin of Species*.

important contributions to the solution of the
problem of biological evolution, Mr. Darwin con-
fines himself to the discussion of the causes which
have brought about the present condition of living
matter, assuming such matter to have once come
into existence. On the other hand, Mr. Spencer [1]
and Professor Haeckel [2] have dealt with the whole
problem of evolution. The profound and vigorous
writings of Mr. Spencer embody the spirit of
Descartes in the knowledge of our own day, and
may be regarded as the "Principes de la
Philosophie" of the nineteenth century; while,
whatever hesitation may not unfrequently be felt
by less daring minds, in following Haeckel in many
of his speculations, his attempt to systematise the
doctrine of evolution and to exhibit its influence
as the central thought of modern biology, cannot
fail to have a far-reaching influence on the progress
of science.

If we seek for the reason of the difference
between the scientific position of the doctrine of
evolution a century ago, and that which it occupies
now, we shall find it in the great accumulation
of facts, the several classes of which have been
enumerated above, under the second to the eighth
heads. For those which are grouped under the
second to the seventh of these classes, respectively,
have a clear significance on the hypothesis of

---

[1] *First Principles,* and *Principles of Biology,* 1860–1864.
[2] *Generelle Morphologie,* 1866.

evolution, while they are unintelligible if that hypothesis be denied. And those of the eighth group are not only unintelligible without the assumption of evolution, but can be proved never to be discordant with that hypothesis, while, in some cases, they are exactly such as the hypothesis requires. The demonstration of these assertions would require a volume, but the general nature of the evidence on which they rest may be briefly indicated.

2. The accurate investigation of the lowest forms of animal life, commenced by Leeuwenhoek and Swammerdam, and continued by the remarkable labours of Reaumur, Trembley, Bonnet, and a host of other observers, in the latter part of the seventeenth and the first half of the eighteenth centuries, drew the attention of biologists to the gradation in the complexity of organisation which is presented by living beings, and culminated in the doctrine of the " échelle des êtres," so powerfully and clearly stated by Bonnet; and, before him, adumbrated by Locke and by Leibnitz. In the then state of knowledge, it appeared that all the species of animals and plants could be arranged in one series; in such a manner that, by insensible gradations, the mineral passed into the plant, the plant into the polype, the polype into the worm, and so, through gradually higher forms of life, to man, at the summit of the animated world.

The conclusions enunciated by Cuvier and Von Baer have been confirmed, in principle, by all subsequent research into the structure of animals and plants. But the effect of the adoption of these conclusions has been rather to substitute a new metaphor for that of Bonnet than to abolish the conception expressed by it. Instead of regarding living things as capable of arrangement in one series like the steps of a ladder, the results of modern investigation compel us to dispose them as if they were the twigs and branches of a tree. The ends of the twigs represent individuals, the smallest groups of twigs species, larger groups genera, and so on, until we arrive at the source of all these ramifications of the main branch, which is represented by a common plan of structure. At the present moment, it is impossible to draw up any definition, based on broad anatomical or developmental characters, by which any one of Cuvier's great groups shall be separated from all the rest. On the contrary, the lower members of each tend to converge towards the lower members of all the others. The same may be said of the vegetable world. The apparently clear distinction between flowering and flowerless plants has been broken down by the series of gradations between the two exhibited by the *Lycopodiaceæ*, *Rhizocarpeæ*, and *Gymnospermeæ*. The groups of *Fungi*, *Lichenes*, and *Algæ* have completely run into one another and, when the lowest forms of each are alone considered, even the animal and vegetable kingdoms cease to have a definite frontier.

If it is permissible to speak of the relations of living forms to one another metaphorically, the similitude chosen must undoubtedly be that of a common root, whence two main trunks, one representing the vegetable and one the animal world, spring; and, each dividing into a few main branches, these subdivide into multitudes of branchlets and these into smaller groups of twigs. As Lamarck has well said—[1]

"Il n'y a que ceux qui se sont longtemps et fortement occupés de la détermination des espèces, et qui ont consulté de riches collections, qui peuvent savoir jusqu'à quel point les *espèces*, parmi les corps vivants se fondent les unes dans les autres, et qui ont pu se convaincre que, dans les parties où nous voyons des *espèces* isolès, cela n'est ainsi que parcequ'il nous en manque d'autres qui en sont plus voisines et que nous n'avons pas encore recueillies.

"Je ne veux pas dire pour cela que les animaux qui existent forment une série très-simple et partout également nuancée ; mais je dis qu'ils forment une série ramense, irréguliérement graduée et qui n'a point de discontinuité dans ses parties, ou qui, du moins, n'en a toujours pas eu, s'il est vrai que, par suite de quelques espèces perdues, il s'en trouve quelque part. Il en resulte que les *espèces* qui terminent chaque rameau de la série générale tiennent, au moins d'un côté, à d'autres espèces voisines qui se nuancent avec elles. Voilà ce que l'état bien connu des choses me met maintenant à portée de demontrer. Je n'ai besoin d'aucune hypothèse ni d'aucune supposition pour cela : j'en atteste tous les naturalistes observateurs."

---

[1] *Philosophie Zoologique,* première partie, chap. iii.

3. In a remarkable essay [1] Meckel remarks—

"There is no good physiologist who has not been struck by the observation that the original form of all organisms is one and the same, and that out of this one form, all, the lowest as well as the highest, are developed in such a manner that the latter pass through the permanent forms of the former as transitory stages. Aristotle, Haller, Harvey, Kielmeyer, Autenrieth, and many others, have either made this observation incidentally, or, especially the latter, have drawn particular attention to it, and deduced therefrom results of permanent importance for physiology."

Meckel proceeds to exemplify the thesis, that the lower forms of animals represent stages in the course of the development of the higher, with a large series of illustrations.

After comparing the Salamanders and the perennibranchiate *Urodela* with the Tadpoles and the Frogs, and enunciating the law that the more highly any animal is organised the more quickly does it pass through the lower stages, Meckel goes on to say—

"From these lowest Vertebrata to the highest, and to the highest forms among these, the comparison between the embryonic conditions of the higher animals and the adult states of the lower can be more completely and thoroughly instituted than if the survey is extended to the Invertebrata, inasmuch as the latter are in many respects constructed upon an altogether too dissimilar type ; indeed they often differ from one another far more than the lowest vertebrate does from the highest mammal ; yet the

---

[1] "Entwurf einer Darstellung der zwischen dem Embryozustände der höheren Thiere und dem permanenten der niederen stattfindenden Parallele," *Beyträge zur Vergleichenden Anatomic,* Bd. ii. 1811.

following pages will show that the comparison may also be
extended to them with interest. In fact, there is a period when,
as Aristotle long ago said, the embryo of the highest animal
has the form of a mere worm ; and, devoid of internal and
external organisation, is merely an almost structureless lump of
polype substance. Nothwithstanding the origin of organs, it
still for a certain time, by reason of its want of an internal bony
skeleton, remains worm and mollusk, and only later enters into
the series of the Vertebrata, although traces of the vertebral
column even in the earliest periods testify its claim to a place
in that series."—*Op. cit.* pp. 4, 5.

If Meckel's proposition is so far qualified, that
the comparison of adult with embryonic forms is
restricted within the limits of one type of organisation ; and, if it is further recollected that the
resemblance between the permanent lower form
and the embryonic stage of a higher form is not
special but general, it is in entire accordance with
modern embryology ; although there is no branch
of biology which has grown so largely, and improved its methods so much, since Meckel's time,
as this.   In its original form, the doctrine of
"arrest of development," as advocated by Geoffroy
Saint-Hilaire and Serres, was no doubt an over-
statement of the case.   It is not true, for example,
that a fish is a reptile arrested in its development,
or that a reptile was ever a fish : but it is true
that the reptile embryo, at one stage of its
development, is an organism which, if it had an
independent existence, must be classified among
fishes ; and all the organs of the reptile pass, in
the course of their development, through conditions

which are closely analogous to those which are permanent in some fishes.

4. That branch of biology which is termed Morphology is a commentary upon, and expansion of, the proposition that widely different animals or plants, and widely different parts of animals or plants, are constructed upon the same plan. From the rough comparison of the skeleton of a bird with that of a man by Belon, in the sixteenth century (to go no farther back), down to the theory of the limbs and the theory of the skull at the present day ; or, from the first demonstration of the homologies of the parts of a flower by C. F. Wolff, to the present elaborate analysis of the floral organs, morphology exhibits a continual advance towards the demonstration of a fundamental unity among the seeming diversities of living structures. And this demonstration has been completed by the final establishment of the cell theory, which involves the admission of a primitive conformity, not only of all the elementary structures in animals and plants respectively, but of those in the one of these great divisions of living things with those in the other. No *à priori* difficulty can be said to stand in the way of evolution, when it can be shown that all animals and all plants proceed by modes of development, which are similar in principle, from a fundamental protoplasmic material.

5. The innumerable cases of structures, which are

rudimentary and apparently useless, in species, the close allies of which possess well-developed and functionally important homologous structures, are readily intelligible on the theory of evolution, while it is hard to conceive their *raison d'être* on any other hypothesis. However, a cautious reasoner will probably rather explain such cases deductively from the doctrine of evolution than endeavour to support the doctrine of evolution by them. For it is almost impossible to prove that any structure, however rudimentary, is useless—that is to say, that it plays no part whatever in the economy; and, if it is in the slightest degree useful, there is no reason why, on the hypothesis of direct creation, it should not have been created. Nevertheless, double-edged as is the argument from rudimentary organs, there is probably none which has produced a greater effect in promoting the general acceptance of the theory of evolution.

6. The older advocates of evolution sought for the causes of the process exclusively in the influence of varying conditions, such as climate and station, or hybridisation, upon living forms. Even Treviranus has got no farther than this point. Lamarck introduced the conception of the action of an animal on itself as a factor in producing modification. Starting from the well-known fact that the habitual use of a limb tends to develop the muscles of the limb, and to produce a greater and greater

facility in using it, he made the general assumption that the effort of an animal to exert an organ in a given direction tends to develop the organ in that direction. But a little consideration showed that, though Lamarck had seized what, as far it goes, is a true cause of modification, it is a cause the actual effects of which are wholly inadequate to account for any considerable modification in animals, and which can have no influence at all in the vegetable world ; and probably nothing contributed so much to discredit evolution, in the early part of this century, as the floods of easy ridicule which were poured upon this part of Lamarck's speculation. The theory of natural selection, or survival of the fittest, was suggested by Wells in 1813, and further elaborated by Matthew in 1831. But the pregnant suggestions of these writers remained practically unnoticed and forgotten, until the theory was independently devised and promulgated by Darwin and Wallace in 1858, and the effect of its publication was immediate and profound.

Those who were unwilling to accept evolution, without better grounds than such as are offered by Lamarck, or the author of that particularly un- satisfactory book, the 'Vestiges of the Natural History of the Creation," and who therefore preferred to suspend their judgment on the question, found, in the principle of selective breeding, pursued in all its applications with marvellous knowledge and skill by Mr. Darwin, a

valid explanation of the occurrence of varieties and
races; and they saw clearly that, if the explanation
would apply to species, it would not only solve the
problem of their evolution, but that it would account for the facts of teleology, as well as for those
of morphology; and for the persistence of some
forms of life unchanged through long epochs of
time, while others undergo comparatively rapid
metamorphosis.

How far "natural selection" suffices for the production of species remains to be seen. Few can
doubt that, if not the whole cause, it is a very important factor in that operation; and that it must
play a great part in the sorting out of varieties
into those which are transitory and those which
are permanent.

But the causes and conditions of variation have
yet to be thoroughly explored; and the importance
of natural selection will not be impaired, even if
further inquiries should prove that variability
is definite, and is determined in certain directions
rather than in others, by conditions inherent in
that which varies. It is quite conceivable that
every species tends to produce varieties of a
limited number and kind, and that the effect of
natural selection is to favour the development of
some of these, while it opposes the development
of others along their predetermined lines of modification.

7. No truths brought to light by biological

investigation were better calculated to inspire
distrust of the dogmas intruded upon science in
the name of theology, than those which relate to
the distribution of animals and plants on the
surface of the earth. Very skilful accommodation
was needful, if the limitation of sloths to South
America, and of the ornithorhynchus to Australia,
was to be reconciled with the literal interpretation
of the history of the deluge; and with the establishment of the existence of distinct provinces of
distribution, any serious belief in the peopling of
the world by migration from Mount Ararat came
to an end.

Under these circumstances, only one alternative
was left for those who denied the occurrence of
evolution—namely, the supposition that the
characteristic animals and plants of each great
province were created as such, within the limits in
which we find them. And as the hypothesis of
"specific centres," thus formulated, was heterodox
from the theological point of view, and unintelligible under its scientific aspect, it may be passed
over without further notice, as a phase of transition from the creational to the evolutional hypothesis.

8. In fact, the strongest and most conclusive
arguments in favour of evolution are those which
are based upon the facts of geographical, taken
in conjunction with those of geological, distribution.

Both Mr. Darwin and Mr. Wallace lay great stress on the close relation which obtains between the existing fauna of any region and that of the immediately antecedent geological epoch in the same region; and rightly, for it is in truth inconceivable that there should be no genetic connection between the two. It is possible to put into words the proposition that all the animals and plants of each geological epoch were annihilated and that a new set of very similar forms was created for the next epoch; but it may be doubted if any one who ever tried to form a distinct mental image of this process of spontaneous generation on the grandest scale, ever really succeeded in realising it.

Within the last twenty years, the attention of the best palæontologists has been withdrawn from the hodman's work of making "new species" of fossils, to the scientific task of completing our knowledge of individual species, and tracing out the succession of the forms presented by any given type in time.

Those who desire to inform themselves of the nature and extent of the evidence bearing on these questions may consult the works of Rütimeyer, Gaudry, Kowalewsky, Marsh, and the writer of the present article. It must suffice, in this place, to say that the successive forms of the Equine type have been fully worked out; while those of nearly all the other existing types of Ungulate mammals and of the *Carnivora* have been almost as closely followed through the Tertiary deposits; the gradations between birds and reptiles have been traced; and the modifications undergone by the *Crocodilia*, from the Triassic epoch to the present day, have been demonstrated. On the evidence of palæontology, the evolution of many existing forms of animal life from their predecessors is no longer an hypothesis, but an historical fact; it is only the nature of the physiological factors to which that evolution is due which is still open to discussion.

[At page 209, the reference to Erasmus Darwin does not do justice to that ingenious writer, who, in the 39th section of the *Zoonomia*, clearly and repeatedly enunciates the theory of the inheritance of acquired modifications. For example: "From their first rudiment, or primordium, to the termination of their lives, all animals undergo perpetual transformations; which are in part produced by their own exertions in consequence of their desires and aversions, of their pleasures and their pains, or of irritation, or of associations; and many of these acquired forms or propensities are transmitted to their posterity." *Zoonomia* I., p. 506. 1893.]

VII

THE COMING OF AGE OF "THE ORIGIN OF SPECIES"

[1880]

MANY of you will be familiar with the aspect of this small green-covered book. It is a copy of the first edition of the "Origin of Species," and bears the date of its production—the 1st of October 1859. Only a few months, therefore, are needed to complete the full tale of twenty-one years since its birthday.

Those whose memories carry them back to this time will remember that the infant was remarkably lively, and that a great number of excellent persons mistook its manifestations of a vigorous individuality for mere naughtiness; in fact there was a very pretty turmoil about its cradle. My recollections of the period are particularly vivid; for, having conceived a tender affection for a child of what appeared to me to be such remarkable promise, I acted for some time in the capacity of a

Q 2

sort of under-nurse, and thus came in for my share
of the storms which threatened the very life of
the young creature. For some years it was
undoubtedly warm work; but considering how
exceedingly unpleasant the apparition of the new-
comer must have been to those who did not fall in
love with him at first sight, I think it is to the
credit of our age that the war was not fiercer, and
that the more bitter and unscrupulous forms of
opposition died away as soon as they did.

I speak of this period as of something past and
gone, possessing merely an historical, I had almost
said an antiquarian interest. For, during the
second decade of the existence of the " Origin of
Species," opposition, though by no means dead,
assumed a different aspect. On the part of all
those who had any reason to respect themselves,
it assumed a thoroughly respectful character. By
this time, the dullest began to perceive that the
child was not likely to perish of any congenital
weakness or infantile disorder, but was growing
into a stalwart personage, upon whom mere goody
scoldings and threatenings with the birch-rod
were quite thrown away.

In fact, those who have watched the progress of
science within the last ten years will bear me out
to the full, when I assert that there is no field of
biological inquiry in which the influence of the
" Origin of Species " is not traceable ; the foremost
men of science in every country are either avowed

champions of its leading doctrines, or at any rate
abstain from opposing them ; a host of young and
ardent investigators seek for and find inspiration
and guidance in Mr. Darwin's great work ; and the
general doctrine of evolution, to one side of which
it gives expression, obtains, in the phenomena of
biology, a firm base of operations whence it may
conduct its conquest of the whole realm of Nature.

History warns us, however, that it is the cus-
tomary fate of new truths to begin as heresies and
to end as superstitions ; and, as matters now stand,
it is hardly rash to anticipate that, in another
twenty years, the new generation, educated under
the influences of the present day, will be in danger
of accepting the main doctrines of the " Origin of
Species," with as little reflection, and it may be
with as little justification, as so many of our contemporaries, twenty years ago, rejected them.

Against any such a consummation let us all
devoutly pray ; for the scientific spirit is of more
value than its products, and irrationally held
truths may be more harmful than reasoned errors.
Now the essence of the scientific spirit is criticism.
It tells us that whenever a doctrine claims our
assent we should reply, Take it if you can compel
it.   The struggle for existence holds as much in
the intellectual as in the physical world.   A theory
is a species of thinking, and its right to exist is
coextensive with its power of resisting extinction
by its rivals.

From this point of view, it appears to me that it would be but a poor way of celebrating the Coming of Age of the " Origin of Species," were I merely to dwell upon the facts, undoubted and remarkable as they are, of its far-reaching influence and of the great following of ardent disciples who are occupied in spreading and developing its doctrines. Mere insanities and inanities have before now swollen to portentous size in the course of twenty years. Let us rather ask this prodigious change in opinion to justify itself : let us inquire whether anything has happened since 1859, which will explain, on rational grounds, why so many are worshipping that which they burned, and burning that which they worshipped. It is only in this way that we shall acquire the means of judging whether the movement we have witnessed is a mere eddy of fashion, or truly one with the irreversible current of intellectual progress, and, like it, safe from retrogressive reaction.

Every belief is the product of two factors : the first is the state of the mind to which the evidence in favour of that belief is presented ; and the second is the logical cogency of the evidence itself. In both these respects, the history of biological science during the last twenty years appears to me to afford an ample explanation of the change which has taken place ; and a brief consideration of the salient events of that history will enable us to understand why, if the " Origin of Species " appeared now, it would meet with a very different reception from that which greeted it in 1859.

One-and-twenty years ago, in spite of the work commenced by Hutton and continued with rare skill and patience by Lyell, the dominant view of the past history of the earth was catastrophic. Great and sudden physical revolutions, wholesale creations and extinctions of living beings, were the ordinary machinery of the geological epic brought into fashion by the misapplied genius of Cuvier. It was gravely maintained and taught that the end of every geological epoch was signalised by a cataclysm, by which every living being on the globe was swept away, to be replaced by a brand-new creation when the world returned to quiescence. A scheme of nature which appeared to be modelled on the likeness of a succession of rubbers of whist, at the end of each of which the players upset the table and called for a new pack, did not seem to shock anybody.

I may be wrong, but I doubt if, at the present time, there is a single responsible representative of these opinions left. The progress of scientific geology has elevated the fundamental principle of uniformitarianism, that the explanation of the past is to be sought in the study of the present, into the position of an axiom ; and the wild speculations of the catastrophists, to which we all listened with respect a quarter of a century ago, would hardly find a single patient hearer at the present day. No physical geologist now dreams of seeking, outside the range of known natural causes, for the explanation of anything that happened millions of years ago, any more than he would be . guilty of the like absurdity in regard to current events.

The effect of this change of opinion upon biological speculation is obvious. For, if there have been no periodical general physical catastrophes, what brought about the assumed general extinctions and re-creations of life which are the corresponding biological catastrophes? And, if no such interruptions of the ordinary course of nature have taken place in the organic, any more than in the inorganic, world, what alternative is there to the admission of evolution?

The doctrine of evolution in biology is the necessary result of the logical application of the principles of uniformitarianism to the phenomena of life. Darwin is the natural successor of Hutton and Lyell, and the " Origin of Species " the logical sequence of the " Principles of Geology."

The fundamental doctrine of the " Origin of Species," as of all forms of the theory of evolution applied to biology, is " that the innumerable species, genera, and families of organic beings with which the world is peopled have all descended, each within its own class or group, from common parents, and have all been modified in the course of descent." [1]

[1] *Origin of Species*, ed. 1, p. 457.

And, in view of the facts of geology, it follows
that all living animals and plants "are the lineal
descendants of those which lived long before the
Silurian epoch." [1]

It is an obvious consequence of this theory of
descent with modification, as it is sometimes called,
that all plants and animals, however different they
may now be, must, at one time or other, have been
connected by direct or indirect intermediate gradations, and that the appearance of isolation presented
by various groups of organic beings must be unreal.

No part of Mr. Darwin's work ran more directly
counter to the prepossessions of naturalists twenty
years ago than this. And such prepossessions were
very excusable, for there was undoubtedly a great
deal to be said, at that time, in favour of the fixity
of species and of the existence of great breaks,
which there was no obvious or probable means of
filling up, between various groups of organic beings.

For various reasons, scientific and unscientific,
much had been made of the hiatus between man
and the rest of the higher mammalia, and it is no
wonder that issue was first joined on this part of
the controversy. I have no wish to revive past
and happily forgotten controversies; but I must
state the simple fact that the distinctions in the
cerebral and other characters, which were so hotly
affirmed to separate man from all other animals in
1860, have all been demonstrated to be non-

[1] *Origin of Species*, p. 458.

existent, and that the contrary doctrine is now universally accepted and taught.

But there were other cases in which the wide structural gaps asserted to exist between one group of animals and another were by no means fictitious; and, when such structural breaks were real, Mr. Darwin could account for them only by supposing that the intermediate forms which once existed had become extinct. In a remarkable passage he says—

"We may thus account even for the distinctness of whole classes from each other—for instance, of birds from all other vertebrate animals—by the belief that many animal forms of life have been utterly lost, through which the early progenitors of birds were formerly connected with the early progenitors of the other vertebrate classes." [1]

Adverse criticism made merry over such suggestions as these. Of course it was easy to get out of the difficulty by supposing extinction; but where was the slightest evidence that such intermediate forms between birds and reptiles as the hypothesis required ever existed? And then probably followed a tirade upon this terrible forsaking of the paths of "Baconian induction."

But the progress of knowledge has justified Mr. Darwin to an extent which could hardly have been anticipated. In 1862, the specimen of *Archæopteryx*, which, until the last two or three

[1] *Origin of Species*, p. 431.

years, has remained unique, was discovered; and
it is an animal which, in its feathers and the
greater part of its organisation, is a veritable
bird, while, in other parts, it is as distinctly
reptilian.

In 1868, I had the honour of bringing under
your notice, in this theatre, the results of investigations made, up to that time, into the anatomical
characters of certain ancient reptiles, which
showed the nature of the modifications in virtue
of which the type of the quadrupedal reptile
passed into that of a bipedal bird ; and abundant
confirmatory evidence of the justice of the conclusions which I then laid before you has since
come to light.

In 1875, the discovery of the toothed birds of
the cretaceous formation in North America by
Professor Marsh completed the series of transitional
forms between birds and reptiles, and removed
Mr. Darwin's proposition that "many animal
forms of life have been utterly lost, through
which the early progenitors of birds were
formerly connected with the early progenitors of
the other vertebrate classes," from the region
of hypothesis to that of demonstrable fact.

In 1859, there appeared to be a very sharp
and clear hiatus between vertebrated and inverte-
brated animals, not only in their structure, but,
what was more important, in their development.
I do not think that we even yet know the precise

links of connection between the two; but the investigations of Kowalewsky and others upon the development of *Amphioxus* and of the *Tunicata* prove, beyond a doubt, that the differences which were supposed to constitute a barrier between the two are non-existent. There is no longer any difficulty in understanding how the vertebrate type may have arisen from the invertebrate, though the full proof of the manner in which the transition was actually effected may still be lacking.

Again, in 1859, there appeared to be a no less sharp separation between the two great groups of flowering and flowerless plants. It is only subsequently that the series of remarkable investigations inaugurated by Hofmeister has brought to light the extraordinary and altogether unexpected modifications of the reproductive apparatus in the *Lycopodiaceæ*, the *Rhizocarpeæ*, and the *Gymnospermeæ*, by which the ferns and the mosses are gradually connected with the Phanerogamic division of the vegetable world.

So, again, it is only since 1859 that we have acquired that wealth of knowledge of the lowest forms of life which demonstrates the futility of any attempt to separate the lowest plants from the lowest animals, and shows that the two kingdoms of living nature have a common borderland which belongs to both, or to neither.

Thus it will be observed that the whole ten-

dency of biological investigation, since 1859, has
been in the direction of removing the difficulties
which the apparent breaks in the series created
at that time; and the recognition of gradation
is the first step towards the acceptance of evolution.

As another great factor in bringing about the
change of opinion which has taken place among
naturalists, I count the astonishing progress which
has been made in the study of embryology.
Twenty years ago, not only were we devoid of any
accurate knowledge of the mode of development
of many groups of animals and plants, but the
methods of investigation were rude and imperfect.
At the present time, there is no important group
of organic beings the development of which has
not been carefully studied; and the modern
methods of hardening and section-making enable
the embryologist to determine the nature of the
process, in each case, with a degree of minuteness
and accuracy which is truly astonishing to those
whose memories carry them back to the
beginnings of modern histology. And the results
of these embryological investigations are in complete harmony with the requirements of the
doctrine of evolution. The first beginnings of all
the higher forms of animal life are similar, and
however diverse their adult conditions, they start
from a common foundation. Moreover, the process of development of the animal or the plant

from its primary egg, or germ, is a true process of
evolution—a progress from almost formless to
more or less highly organised matter, in virtue of
the properties inherent in that matter.

To those who are familiar with the process of
development, all à priori objections to the doctrine
of biological evolution appear childish. Any one
who has watched the gradual formation of a complicated animal from the protoplasmic mass, which
constitutes the essential element of a frog's or a
hen's egg, has had under his eyes sufficient
evidence that a similar evolution of the whole
animal world from the like foundation is, at any
rate, possible.

Yet another product of investigation has
largely contributed to the removal of the objections to the doctrine of evolution current in 1859.
It is the proof afforded by successive discoveries
that Mr. Darwin did not over-estimate the
imperfection of the geological record. No more
striking illustration of this is needed than a comparison of our knowledge of the mammalian fauna
of the Tertiary epoch in 1859 with its present
condition. M. Gaudry's researches on the fossils
of Pikermi were published in 1868, those of
Messrs. Leidy, Marsh, and Cope, on the fossils of
the Western Territories of America, have appeared
almost wholly since 1870, those of M. Filhol on
the phosphorites of Quercy in 1878. The general
effect of these investigations has been to intro-

duce to us a multitude of extinct animals, the
existence of which was previously hardly sus-
pected; just as if zoologists were to become
acquainted with a country, hitherto unknown, as
rich in novel forms of life as Brazil or South
Africa once were to Europeans. Indeed, the fossil
fauna of the Western Territories of America bid
fair to exceed in interest and importance all other
known Tertiary deposits put together; and yet,
with the exception of the case of the American
tertiaries, these investigations have extended over
very limited areas; and, at Pikermi, were confined to an extremely small space.

Such appear to me to be the chief events in the
history of the progress of knowledge during the
last twenty years, which account for the changed
feeling with which the doctrine of evolution is at
present regarded by those who have followed the
advance of biological science, in respect of those
problems which bear indirectly upon that doctrine.

But all this remains mere secondary evidence.
It may remove dissent, but it does not compel
assent. Primary and direct evidence in favour of
evolution can be furnished only by palæontology.
The geological record, so soon as it approaches
completeness, must, when properly questioned,
yield either an affirmative or a negative answer:
if evolution has taken place, there will its mark

be left; if it has not taken place, there will lie
its refutation.

What was the state of matters in 1859? Let
us hear Mr. Darwin, who may be trusted always
to state the case against himself as strongly as
possible.

" On this doctrine of the extermination of an
infinitude of connecting links between the living
and extinct inhabitants of the world, and at each
successive period between the extinct and still
older species, why is not every geological formation charged with such links? Why does not
every collection of fossil remains afford plain
evidence of the gradation and mutation of the
forms of life? We meet with no such evidence,
and this is the most obvious and plausible of the
many objections which may be urged against my
theory." [1]

Nothing could have been more useful to the
opposition than this characteristically candid
avowal, twisted as it immediately was into an
admission that the writer's views were contradicted by the facts of palæontology. But, in fact,
Mr. Darwin made no such admission. What he
says in effect is, not that palæontological evidence
is against him, but that it is not distinctly in his
favour; and, without attempting to attenuate the
fact, he accounts for it by the scantiness and the
imperfection of that evidence.

[1] *Origin of Species*, ed. 1, p. 463.

What is the state of the case now, when, as we have seen, the amount of our knowledge respecting the mammalia of the Tertiary epoch is increased fifty-fold, and in some directions even approaches completeness?

Simply this, that, if the doctrine of evolution had not existed, palæontologists must have invented it, so irresistibly is it forced upon the mind by the study of the remains of the Tertiary mammalia which have been brought to light since 1859.

Among the fossils of Pikermi, Gaudry found the successive stages by which the ancient civets passed into the more modern hyænas; through the Tertiary deposits of Western America, Marsh tracked the successive forms by which the ancient stock of the horse has passed into its present form; and innumerable less complete indications of the mode of evolution of other groups of the higher mammalia have been obtained. In the remarkable memoir on the phosphorites of Quercy, to which I have referred, M. Filhol describes no fewer than seventeen varieties of the genus *Cynodictis*, which fill up all the interval between the viverine animals and the bear-like dog *Amphicyon*; nor do I know any solid ground of objection to the supposition that, in this *Cynodictis-Amphicyon* group, we have the stock whence all the Viveridæ, Felidæ, Hyænidæ, Canidæ, and perhaps the Procyonidæ and Ursidæ, of the present fauna have been evolved. On the contrary, there is a great deal to be said in favour.

In the course of summing up his results, M. Filhol observes :—

"During the epoch of the phosphorites, great changes took place in animal forms, and almost the same types as those which now exist became defined from one another.

"Under the influence of natural conditions of which we have no exact knowledge, though traces of them are discoverable, species have been modified in a thousand ways : races have arisen which, becoming fixed, have thus produced a corresponding number of secondary species."

In 1859, language of which this is an unintentional paraphrase, occurring in the "Origin of Species," was scouted as wild speculation; at present, it is a sober statement of the conclusions to which an acute and critically-minded investigator is led by large and patient study of the facts of palæontology. I venture to repeat what I have said before, that so far as the animal world is concerned, evolution is no longer a speculation, but a statement of historical fact. It takes its place alongside of those accepted truths which must be reckoned with by philosophers of all schools.

Thus when, on the first day of October next, "The Origin of Species" comes of age, the promise of its youth will be amply fulfilled ; and we shall be prepared to congratulate the venerated author of the book, not only that the greatness of his achievement and its enduring influence upon the progress of knowledge have won him a place beside our Harvey; but, still more, that, like Harvey, he has lived long enough to outlast detraction and opposition, and to see the stone that the builders rejected become the head-stone of the corner.

VIII

CHARLES DARWIN

[*Nature*, April 27th, 1882]

VERY few, even among those who have taken the keenest interest in the progress of the revolution in natural knowledge set afoot by the publication of " The Origin of Species," and who have watched, not without astonishment, the rapid and complete change which has been effected both inside and outside the boundaries of the scientific world in the attitude of men's minds towards the doctrines which are expounded in that great work, can have been prepared for the extraordinary manifestation of affectionate regard for the man, and of profound reverence for the philosopher, which followed the announcement, on Thursday last, of the death of Mr. Darwin.

Not only in these islands, where so many have felt the fascination of personal contact with an

intellect which had no superior, and with a character which was even nobler than the intellect; but,
in all parts of the civilised world, it would seem
that those whose business it is to feel the pulse of
nations and to know what interests the masses of
mankind, were well aware that thousands of their
readers would think the world the poorer for
Darwin's death, and would dwell with eager
interest upon every incident of his history. In
France, in Germany, in Austro-Hungary, in Italy,
in the United States, writers of all shades of
opinion, for once unanimous, have paid a willing
tribute to the worth of our great countryman,
ignored in life by the official representatives of the
kingdom, but laid in death among his peers in
Westminster Abbey by the will of the intelligence
of the nation.

It is not for us to allude to the sacred sorrows
of the bereaved home at Down; but it is no secret
that, outside that domestic group, there are many
to whom Mr. Darwin's death is a wholly irreparable
loss. And this not merely because of his wonder-
fully genial, simple, and generous nature; his
cheerful and animated conversation, and the infinite variety and accuracy of his information; but
because the more one knew of him, the more he
seemed the incorporated ideal of a man of science.
Acute as were his reasoning powers, vast as was
his knowledge, marvellous as was his tenacious
industry, under physical difficulties which would

have converted nine men out of ten into aimless invalids; it was not these qualities, great as they were, which impressed those who were admitted to his intimacy with involuntary veneration, but a certain intense and almost passionate honesty by which all his thoughts and actions were irradiated, as by a central fire.

It was this rarest and greatest of endowments which kept his vivid imagination and great speculative powers within due bounds; which compelled him to undertake the prodigious labours of original investigation and of reading, upon which his published works are based; which made him accept criticisms and suggestions from anybody and everybody, not only without impatience, but with expressions of gratitude sometimes almost comically in excess of their value; which led him to allow neither himself nor others to be deceived by phrases, and to spare neither time nor pains in order to obtain clear and distinct ideas upon every topic with which he occupied himself.

One could not converse with Darwin without being reminded of Socrates. There was the same desire to find some one wiser than himself; the same belief in the sovereignty of reason; the same ready humour; the same sympathetic interest in all the ways and works of men. But instead of turning away from the problems of Nature as hopelessly insoluble, our modern philosopher devoted his whole life to attacking them in the spirit of Heraclitus and of Democritus, with results which are the substance of which their speculations were anticipatory shadows.

The due appreciation, or even enumeration, of these results is neither practicable nor desirable at this moment. There is a time for all things—a time for glorying in our ever-extending conquests over the realm of Nature, and a time for mourning over the heroes who have led us to victory.

None have fought better, and none have been more fortunate, than Charles Darwin. He found a great truth trodden underfoot, reviled by bigots, and ridiculed by all the world; he lived long enough to see it, chiefly by his own efforts, irrefragably established in science, inseparably incorporated with the common thoughts of men, and only hated and feared by those who would revile, but dare not. What shall a man desire more than this? Once more the image of Socrates rises unbidden, and the noble peroration of the "Apology" rings in our ears as if it were Charles Darwin's farewell :—

"The hour of departure has arrived, and we go our ways—I to die and you to live. Which is the better, God only knows."

IX

THE DARWIN MEMORIAL

[June 9th, 1885]

Address by the President of the Royal Society, in the name of the Memorial Committee, on handing over the statue of Darwin to H.R.H. the Prince of Wales, as representative of the Trustees of the British Museum.

YOUR ROYAL HIGHNESS,—It is now three years since the announcement of the death of our famous countryman, Charles Darwin, gave rise to a manifestation of public feeling, not only in these realms, but throughout the civilised world, which, if I mistake not, is without precedent in the modest annals of scientific biography.

The causes of this deep and wide outburst of emotion are not far to seek. We had lost one of these rare ministers and interpreters of Nature whose names mark epochs in the advance of

natural knowledge. For, whatever be the ultimate
verdict of posterity upon this or that opinion
which Mr. Darwin has propounded ; whatever
adumbrations or anticipations of his doctrines may
be found in the writings of his predecessors ; the
broad fact remains that, since the publication and
by reason of the publication, of " The Origin of
Species " the fundamental conceptions and the
aims of the students of living Nature have been
completely changed. From that work has sprung
a great renewal, a true " instauratio magna " of the
zoological and botanical sciences.

But the impulse thus given to scientific thought
rapidly spread beyond the ordinarily recognised
limits of biology. Psychology, Ethics, Cosmology
were stirred to their foundations, and the " Origin
of Species " proved itself to be the fixed point
which the general doctrine of evolution needed in
order to move the world. " Darwinism," in one
form or another, sometimes strangely distorted
and mutilated, became an everyday topic of men's
speech, the object of an abundance both of
vituperation and of praise, more often than of
serious study.

It is curious now to remember how largely, at
first, the objectors predominated ; but considering
the usual fate of new views, it is still more
curious to consider for how short a time the phase
of vehement opposition lasted. Before twenty
years had passed, not only had the importance of

Mr. Darwin's work been fully recognised, but the world had discerned the simple, earnest, generous character of the man, that shone through every page of his writings.

I imagine that reflections such as these swept through the minds alike of loving friends and of honourable antagonists when Mr. Darwin died; and that they were at one in the desire to honour the memory of the man who, without fear and without reproach, had successfully fought the hardest intellectual battle of these days.

It was in satisfaction of these just and generous impulses that our great naturalist's remains were deposited in Westminster Abbey; and that, immediately afterwards, a public meeting, presided over by my lamented predecessor, Mr. Spottiswoode, was held in the rooms of the Royal Society, for the purpose of considering what further step should be taken towards the same end.

It was resolved to invite subscriptions, with the view of erecting a statue of Mr. Darwin in some suitable locality; and to devote any surplus to the advancement of the biological sciences.

Contributions at once flowed in from Austria, Belgium, Brazil, Denmark, France, Germany, Holland, Italy, Norway, Portugal, Russia, Spain, Sweden, Switzerland, the United States, and the British Colonies, no less than from all parts of the three kingdoms; and they came from all classes of the community. To mention one interesting case,

Sweden sent in 2296 subscriptions "from all sorts of people," as the distinguished man of science who transmitted them wrote, "from the bishop to the seamstress, and in sums from five pounds to two pence."

The Executive Committee has thus been enabled to carry out the objects proposed. A "Darwin Fund" has been created, which is to be held in trust by the Royal Society, and is to be employed in the promotion of biological research.

The execution of the statue was entrusted to Mr. Boehm; and I think that those who had the good fortune to know Mr. Darwin personally will admire the power of artistic divination which has enabled the sculptor to place before us so very characteristic a likeness of one whom he had not seen.

It appeared to the Committee that, whether they regarded Mr. Darwin's career or the requirements of a work of art, no site could be so appropriate as this great hall, and they applied to the Trustees of the British Museum for permission to erect it in its present position.

That permission was most cordially granted, and I am desired to tender the best thanks of the Committee to the Trustees for their willingness to accede to our wishes.

I also beg leave to offer the expression of our gratitude to your Royal Highness for kindly consenting to represent the Trustees to-day.

It only remains for me, your Royal Highness, my Lords and Gentlemen, Trustees of the British Museum, in the name of the Darwin Memorial Committee, to request you to accept this statue of Charles Darwin.

We do not make this request for the mere sake of perpetuating a memory ; for so long as men occupy themselves with the pursuit of truth, the name of Darwin runs no more risk of oblivion than does that of Copernicus, or that of Harvey.

Nor, most assuredly, do we ask you to preserve the statue in its cynosural position in this entrance-hall of our National Museum of Natural History as evidence that Mr. Darwin's views have received your official sanction ; for science does not recognise such sanctions, and commits suicide when it adopts a creed.

No ; we beg you to cherish this Memorial as a symbol by which, as generation after generation of students of Nature enter yonder door, they shall be reminded of the ideal according to which they must shape their lives, if they would turn to the best account the opportunities offered by the great institution under your charge.

X

OBITUARY [1]

[1888]

CHARLES ROBERT DARWIN was the fifth child
and second son of Robert Waring Darwin and
Susannah Wedgwood, and was born on the 12th
February, 1809, at Shrewsbury, where his father
was a physician in large practice.

Mrs. Robert Darwin died when her son Charles
was only eight years old, and he hardly remem-
bered her. A daughter of the famous Josiah
Wedgwood, who created a new branch of the
potter's art, and established the great works of
Etruria, could hardly fail to transmit important
mental and moral qualities to her children; and
there is a solitary record of her direct influence
in the story told by a schoolfellow, who remembers
Charles Darwin "bringing a flower to school, and

[1] From the Obituary Notices of the *Proceedings of the Royal Society*, vol. 44.

saying that his mother had taught him how, by looking at the inside of the blossom, the name of the plant could be discovered." (I., p. 28.[1])

The theory that men of genius derive their qualities from their mothers, however, can hardly derive support from Charles Darwin's case, in the face of the patent influence of his paternal fore-fathers. Dr. Darwin, indeed, though a man of marked individuality of character, a quick and acute observer, with much practical sagacity, is said not to have had a scientific mind. But when his son adds that his father " formed a theory for almost everything that occurred " (I., p. 20), he indicates a highly probable source for that in-ability to refrain from forming an hypothesis on every subject which he confesses to be one of the leading characteristics of his own mind, some pages further on (I., p. 103). Dr. R. W. Darwin, again, was the third son of Erasmus Darwin, also a physician of great repute, who shared the intimacy of Watt and Priestley, and was widely known as the author of " Zoonomia," and other voluminous poetical and prose works which had a great vogue in the latter half of the eighteenth century. The celebrity which they enjoyed was in part due to the attractive style (at least according to the taste of that day) in which the author's extensive, though not very profound,

---

[1] The references throughout this notice are to the *Life and Letters*, unless the contrary is expressly stated.

acquaintance with natural phenomena was set
forth; but in a still greater degree, probably, to
the boldness of the speculative views, always
ingenious and sometimes fantastic, in which he
indulged. The conception of evolution set afoot
by De Maillet and others, in the early part of the
century, not only found a vigorous champion in
Erasmus Darwin, but he propounded an hypothesis as to the manner in which the species of
animals and plants have acquired their characters,
which is identical in principle with that subsequently rendered famous by Lamarck.

That Charles Darwin's chief intellectual inheritance came to him from the paternal side,
then, is hardly doubtful. But there is nothing to
show that he was, to any sensible extent, directly
influenced by his grandfather's biological work.
He tells us that a perusal of the "Zoonomia" in
early life produced no effect upon him, although
he greatly admired it; and that, on reading it again,
ten or fifteen years afterwards, he was much disappointed, "the proportion of speculation being so
large to the facts given." But with his usual
anxious candour he adds, "Nevertheless, it is probable that the hearing, rather early in life, such views
maintained and praised, may have favoured my
upholding them, in a different form, in my Origin
of Species.'" (I., p. 38.) Erasmus Darwin was in
fact an anticipator of Lamarck, and not of Charles
Darwin; there is no trace in his works of the

conceptions by the addition of which his grandson metamorphosed the theory of evolution as applied to living things and gave it a new foundation.

Charles Darwin's childhood and youth afforded no intimation that he would be, or do, anything out of the common run. In fact, the prognostications of the educational authorities into whose hands he first fell were most distinctly unfavourable; and they counted the only boy of original genius who is known to have come under their hands as no better than a dunce. The history of the educational experiments to which Darwin was subjected is curious, and not without a moral for the present generation. There were four of them, and three were failures. Yet it cannot be said that the materials on which the pedagogic powers operated were other than good. In his boyhood Darwin was strong, well-grown, and active, taking the keen delight in field sports and in every description of hard physical exercise which is natural to an English country-bred lad; and, in respect of things of the mind, he was neither apathetic, nor idle, nor one-sided. The "Autobiography" tells us that he "had much zeal for whatever interested" him, and he was interested in many and very diverse topics. He could work hard, and liked a complex subject better than an easy one. The "clear geometrical proofs" of Euclid delighted him. His interest in practical chemistry, carried out in

an extemporised laboratory, in which he was permitted to assist by his elder brother, kept him
late at work, and earned him the nickname of
" gas " among his schoolfellows. And there could
have been no insensibility to literature in one
who, as a boy, could sit for hours reading Shakespeare, Milton, Scott, and Byron; who greatly
admired some of the Odes of Horace; and who,
in later years, on board the " Beagle," when only
one book could be carried on an expedition,
chose a volume of Milton for his companion.

Industry, intellectual interests, the capacity for
taking pleasure in deductive reasoning, in observation, in experiment, no less than in the highest
works of imagination : where these qualities are
present any rational system of education should
surely be able to make something of them. Unfortunately for Darwin, the Shrewsbury Grammar
School, though good of its kind, was an institution
of a type universally prevalent in this country half
a century ago, and by no means extinct at the
present day. The education given was "strictly
classical," "especial attention" being "paid to
verse-making," while all other subjects, except a
little ancient geography and history, were ignored.
Whether, as in some famous English schools at that
date and much later, elementary arithmetic was
also left out of sight does not appear; but the
instruction in Euclid which gave Charles Darwin
so much satisfaction was certainly supplied by a

ticular aptitude for grammatical exercises; appeared to the "strictly classical" pedagogue to be no mind at all. As a matter of fact, Darwin's school education left him ignorant of almost all the things which it would have been well for him to know, and untrained in all the things it would have been useful for him to be able to do, in after life. Drawing, practice in English composition, and instruction in the elements of the physical sciences, would not only have been infinitely valuable to him in reference to his future career, but would have furnished the discipline suited to his faculties, whatever that career might be. And a knowledge of French and German, especially the latter, would have removed from his path obstacles which he never fully overcame.

Thus, starved and stunted on the intellectual side, it is not surprising that Charles Darwin's energies were directed towards athletic amusements and sport, to such an extent, that even his kind and sagacious father could be exasperated into telling him that "he cared for nothing but shooting, dogs, and rat-catching." (I. p. 32.) It would be unfair to expect even the wisest of fathers to have foreseen that the shooting and the rat-catching, as training in the ways of quick observation and in physical endurance, would prove more valuable than the construing and verse-making to his son, whose attempt, at a later period of his life, to persuade himself "that shooting was almost an

s 2

intellectual employment : it required so much skill
to judge where to find most game, and to hunt the
dogs well " (I. p. 43), was by no means so sophis-
tical as he seems to have been ready to admit.

In 1825, Dr. Darwin came to the very just conclusion that his son Charles would do no good by
remaining at Shrewsbury School, and sent him to
join his elder brother Erasmus, who was studying
medicine at Edinburgh, with the intention that
the younger son should also become a medical
practitioner. Both sons, however, were well aware
that their inheritance would relieve them from the
urgency of the struggle for existence which most
professional men have to face ; and they seemed to
have allowed their tastes, rather than the medical
curriculum, to have guided their studies. Erasmus
Darwin was debarred by constant ill-health from
seeking the public distinction which his high intelligence and extensive knowledge would, under
ordinary circumstances, have insured. He took
no great interest in biological subjects,' but his
companionship must have had its influence on
his brother. Still more was exerted by friends
like Coldstream and Grant, both subsequently
well-known zoologists (and the latter an enthusiastic Lamarckian), by whom Darwin was induced
to interest himself in marine zoology. A notice
of the ciliated germs of *Flustra*, communicated to
the Plinian Society in 1826, was the first fruits of
Darwin's half century of scientific work. Occa-

sional attendance at the Wernerian Society brought
him into relation with that excellent ornithologist
the elder Macgillivray, and enabled him to see and
hear Audubon. Moreover, he got lessons in bird-
stuffing from a negro, who had accompanied the
eccentric traveller Waterton in his wanderings,
before settling in Edinburgh.

No doubt Darwin picked up a great deal of
valuable knowledge during his two years' residence
in Scotland; but it is equally clear that next to
none of it came through the regular channels of
academic education. Indeed, the influence of the
Edinburgh professoriate appears to have been
mainly negative, and in some cases deterrent;
creating in his mind, not only a very low estimate
of the value of lectures, but an antipathy to the
subjects which had been the occasion of the
boredom inflicted upon him by their instrument-
ality. With the exception of Hope, the Professor
of Chemistry, Darwin found them all "intolerably
dull." Forty years afterwards he writes of the
lectures of the Professor of Materia Medica that
they were "fearful to remember." The Professor
of Anatomy made his lectures "as dull as he was
himself," and he must have been very dull to have
wrung from his victim the sharpest personal remark
recorded as his. But the climax seems to have
been attained by the Professor of Geology and
Zoology, whose prælections were so "incredibly
dull" that they produced in their hearer the some-

what rash determination never "to read a book on geology or in any way to study the science" so long as he lived. (I. p. 41.)

There is much reason to believe that the lectures in question were eminently qualified to produce the impression which they made; and there can be little doubt, that Darwin's conclusion that his time was better employed in reading than in listening to such lectures was a sound one. But it was particularly unfortunate that the personal and professorial dulness of the Professor of Anatomy, combined with Darwin's sensitiveness to the disagreeable concomitants of anatomical work, drove him away from the dissecting room. In after life, he justly recognised that this was an "irremediable evil" in reference to the pursuits he eventually adopted; indeed, it is marvellous that he succeeded in making up for his lack of anatomical discipline, so far as his work on the Cirripedes shows he did. And the neglect of anatomy had the further unfortunate result that it excluded him from the best opportunity of bringing himself into direct contact with the facts of nature which the University had to offer. In those days, almost the only practical scientific work accessible to students was anatomical, and the only laboratory at their disposal the dissecting room.

We may now console ourselves with the reflection that the partial evil was the general

good. Darwin had already shown an aptitude for
practical medicine (I. p. 37); and his subsequent
career proved that he had the making of an
excellent anatomist. Thus, though his horror of
operations would probably have shut him off from
surgery, there was nothing to prevent him (any
more than the same peculiarity prevented his
father) from passing successfully through the
medical curriculum and becoming, like his father
and grandfather, a successful physician, in which
case " The Origin of Species " would not have been
written. Darwin has jestingly alluded to the
fact that the shape of his nose (to which Captain
Fitzroy objected), nearly prevented his embarka-
tion in the "Beagle"; it may be that the
sensitiveness of that organ secured him for
science.

At the end of two years' residence in Edinburgh it hardly needed Dr. Darwin's sagacity to
conclude that a young man, who found nothing
but dulness in professorial lucubrations, could not
bring himself to endure a dissecting room, fled
from operations, and did not need a profession as
a means of livelihood, was hardly likely to
distinguish himself as a student of medicine. He
therefore made a new suggestion, proposing that
his son should enter an English University and
qualify for the ministry of the Church. Charles
Darwin found the proposal agreeable, none the
less, probably, that a good deal of natural history

and a little shooting were by no means held, at that time, to be incompatible with the conscientious performance of the duties of a country clergyman. But it is characteristic of the man, that he asked time for consideration, in order that he might satisfy himself that he could sign the Thirty-nine Articles with a clear conscience. However, the study of "Pearson on the Creeds" and a few other books of divinity soon assured him that his religious opinions left nothing to be desired on the score of orthodoxy, and he acceded to his father's proposition.

The English University selected was Cambridge; but an unexpected obstacle arose from the fact that, within the two years which had elapsed, since the young man who had enjoyed seven years of the benefit of a strictly classical education had left school, he had forgotten almost everything he had learned there, "even to some few of the Greek letters." (I. p. 46.) Three months with a tutor, however, brought him back to the point of translating Homer and the Greek Testament "with moderate facility," and Charles Darwin commenced the third educational experiment of which he was the subject, and was entered on the books of Christ's College in October 1827. So far as the direct results of the academic training thus received are concerned, the English University was not more successful than the Scottish. "During the three years which I spent at Cambridge my time was wasted, as far as the academical studies were concerned, as completely as at Edinburgh and as at school." (I. p. 46.) And yet, as before, there is ample evidence that this negative result cannot be put down to any native defect on the part of the scholar. Idle and dull young men, or even young men who being neither idle nor dull, are incapable of caring for anything but some hobby, do not devote themselves to the thorough study of Paley's "Moral Philosophy," and "Evidences of Christianity"; nor are their reminiscences of this particular portion of their studies expressed in terms such as the following : "The logic of this book [the 'Evidences'] and, as I may add, of his 'Natural Theology' gave me as much delight as did Euclid." (I. p. 47.)

The collector's instinct, strong in Darwin from his childhood, as is usually the case in great naturalists, turned itself in the direction of Insects during his residence at Cambridge. In childhood it had been damped by the moral scruples of a sister, as to the propriety of catching and killing insects for the mere sake of possessing them, but now it broke out afresh, and Darwin became an enthusiastic beetle collector. Oddly enough he took no scientific interest in beetles, not even troubling himself to make out their names ; his delight lay in the capture of a species which turned out to be rare or new, and still more in finding his name, as captor, recorded in print. Evidently, this beetle-hunting hobby had little to do with science, but was mainly a new phase of the old and undiminished love of sport. In the intervals of beetle-catching, when shooting and hunting were not to be had, riding across country answered the purpose. These tastes naturally threw the young undergraduate among a set of men who preferred hard riding to hard reading, and wasted the midnight oil upon other pursuits than that of academic distinction. A superficial observer might have had some grounds to fear that Dr. Darwin's wrathful prognosis might yet be verified. But if the eminently social tendencies of a vigorous and genial nature sought an outlet among a set of jovial sporting friends, there were other and no less strong proclivities which brought him into relation with associates of a very different stamp.

Though almost without ear and with a very defective memory for music, Darwin was so strongly and pleasurably affected by it that he became a member of a musical society; and an equal lack of natural capacity for drawing did not prevent him from studying good works of art with much care.

An acquaintance with even the rudiments of physical science was no part of the requirements for the ordinary Cambridge degree. But there were professors both of Geology and of Botany

whose lectures were accessible to those who chose
to attend them. The occupants of these chairs, in
Darwin's time, were eminent men and also admirable lecturers in their widely different styles. The
horror of geological lectures which Darwin had
acquired at Edinburgh, unfortunately prevented
him from going within reach of the fervid elo-
quence of Sedgwick ; but he attended the botanical
course, and though he paid no serious attention to
the subject, he took great delight in the country
excursions, which Henslow so well knew how to
make both pleasant and instructive. The
Botanical Professor was, in fact, a man of rare
character and singularly extensive acquirements
in all branches of natural history. It was his
greatest pleasure to place his stores of knowledge
at the disposal of the young men who gathered
about him, and who found in him, not merely an
encyclopedic teacher but a wise counsellor, and,
in case of worthiness, a warm friend. Darwin's
acquaintance with him soon ripened into a friendship which was terminated only by Henslow's
death in 1861, when his quondam pupil gave
touching expression to his sense of what he owed
to one whom he calls (in one of his letters) his
" dear old master in Natural History." (II. p. 217.)
It was by Henslow's advice that Darwin was led
to break the vow he had registered against making
an acquaintance with geology ; and it was through
Henslow's good offices with Sedgwick that he

seems to have grown very shadowy. Humboldt's "Personal Narrative," and Herschel's "Introduction to the Study of Natural Philosophy," fell in his way and revealed to him his real vocation. The impression made by the former work was very strong. "My whole course of life," says Darwin in sending a message to Humboldt, "is due to having read and re-read, as a youth, his personal narrative." (I. p. 336.) The description of Teneriffe inspired Darwin with such a strong desire to visit the island, that he took some steps towards going there—inquiring about ships, and so on.

But, while this project was fermenting, Henslow, who had been asked to recommend a naturalist for Captain Fitzroy's projected expedition, at once thought of his pupil. In his letter of the 24th August, 1831, he says: "I have stated that I consider you to be the best qualified person I know of who is likely to undertake such a situation. I state this—not on the supposition of your being a *finished* naturalist, but as amply qualified for collecting, observing, and noting anything worthy to be noted in Natural History . . . . The voyage is to last two years, and if you take plenty of books with you, anything you please may be done." (I. p. 193.) The state of the case could not have been better put. Assuredly the young naturalist's theoretical and practical scientific training had gone no further than might suffice for the outfit

of an intelligent collector and note-taker   He was
fully conscious of the fact, and his ambition hardly
rose above the hope that he should bring back
materials for the scientific "lions" at home of
sufficient excellence to prevent them from turning
and rending him. (I. p. 248.)

But a fourth educational experiment was to be
tried. This time Nature took him in hand herself
and showed him the way by which, to borrow
Henslow's prophetic phrase, "anything he pleased
might be done."

The conditions of life presented by a ship-of-war
of only 242 tons burthen, would not, *primâ facie*,
appear to be so favourable to intellectual development as those offered by the cloistered retirement
of Christ's College.   Darwin had not even a cabin
to himself; while, in addition to the hindrances
and interruptions incidental to sea-life, which can
be appreciated only by those who have had
experience of them, sea-sickness came on whenever
the little ship was "lively"; and, considering the
circumstances of the cruise, that must have been
her normal state.   Nevertheless, Darwin found on
board the "Beagle" that which neither the
pedagogues of Shrewsbury, nor the professoriate
of Edinburgh, nor the tutors of Cambridge had
managed to give him.   "I have always felt that I
owe to the voyage the first real training or
education of my mind (I. p. 61); " and in a letter
written as he was leaving England, he calls the

voyage on which he was starting, with just insight, his "second life." (I. p. 214.) Happily for Darwin's education, the school time of the "Beagle" lasted five years instead of two; and the countries which the ship visited were singularly well fitted to provide him with object-lessons, on the nature of things, of the greatest value.

While at sea, he diligently collected, studied, and made copious notes upon the surface Fauna. But with no previous training in dissection, hardly any power of drawing, and next to no knowledge of comparative anatomy, his occupation with work of this kind—notwithstanding all his zeal and industry—resulted, for the most part, in a vast accumulation of useless manuscript. Some acquaintance with the marine *Crustacea*, observations on *Planariæ* and on the ubiquitous *Sagitta*, seem to have been the chief results of a great amount of labour in this direction.

It was otherwise with the terrestrial phenomena which came under the voyager's notice : and Geology very soon took her revenge for the scorn which the much-bored Edinburgh student had poured upon her. Three weeks after leaving England the ship touched land for the first time at St. Jago, in the Cape de Verd Islands, and Darwin found his attention vividly engaged by the volcanic phenomena and the signs of upheaval which the island presented. His geological studies had already indicated the direction in

which a great deal might be done, beyond collecting; and it was while sitting beneath a low lava
cliff on the shore of this island, that a sense of his
real capability first dawned upon Darwin, and
prompted the ambition to write a book on the
geology of the various countries visited. (I. p. 66.)
Even at this early date, Darwin must have thought
much on geological topics, for he was already
convinced of the superiority of Lyell's views to
those entertained by the catastrophists [1]; and his
subsequent study of the tertiary deposits and of the
terraced gravel beds of South America was
eminently fitted to strengthen that conviction.
The letters from South America contain little
reference to any scientific topic except geology;
and even the theory of the formation of coral
reefs was prompted by the evidence of extensive
and gradual changes of level afforded by the
geology of South America; "No other work of
mine," he says, "was begun in so deductive a spirit
as this; for the whole theory was thought out on
the West Coast of South America, before I had
seen a true coral reef. I had, therefore, only to
verify and extend my views by a careful exam-

[1] " I had brought with me the first volume of Lyell's *Principles
of Geology*, which I studied attentively; and the book was of
the highest service to me in many ways. The very first place
which I examined, namely, St. Jago, in the Cape de Verd
Islands, showed me clearly the wonderful superiority of Lyell's
manner of treating Geology, compared with that of any other
author whose works I had with me or ever afterwards
read "—(I. p. 62.)

ination of living reefs.' (I. p. 70.) In 1835, when starting from Lima for the Galapagos, he recommends his friend, W. D. Fox, to take up geology: —" There is so much larger a field for thought than in the other branches of Natural History. I am become a zealous disciple of Mr. Lyell's views, as made known in his admirable book. Geologising in South America, I am tempted to carry parts to a greater extent even than he does. Geology is a capital science to begin with, as it requires nothing but a little reading, thinking, and hammering." (I. p. 263.) The truth of the last statement, when it was written, is a curious mark of the subsequent progress of geology. Even so late as 1836, Darwin speaks of being " much more inclined for geology than the other branches of Natural History." (I. p. 275.)

At the end of the letter to Mr. Fox, however, a little doubt is expressed whether zoological studies might not, after all, have been more profitable ; and an interesting passage in the " Autobiography " enables us to understand the origin of this hesitation.

" During the voyage of the ' Beagle ' I had been deeply impressed by discovering in the Pampean formation great fossil animals covered with armour like that on the existing armadillos ; secondly, by the manner in which closely-allied animals replace one another in proceeding southwards over the continent ; and, thirdly, by the South American

character of most of the productions of the
Galapagos Archipelago, and, more especially, by
the manner in which they differ slightly on each
island of the group; some of the islands appearing
to be very ancient in a geological sense.

"It was evident that such facts as these, as well
as many others, could only be explained on the
supposition that species gradually become modified; and the subject haunted me. But it was
equally evident that neither the action of the
surrounding conditions, nor the will of the organisms (especially in the case of plants) could account
for the innumerable cases in which organisms of
every kind are beautifully adapted to their habits
of life; for instance, a woodpecker or a tree-frog to
climb trees, or a seed for dispersal by hooks or
plumes. I had always been much struck by such
adaptations, and until these could be explained it
seemed to me almost useless to endeavour to prove
by indirect evidence that species have been modified." (I. p. 82.)

The facts to which reference is here made were,
without doubt, eminently fitted to attract the attention of a philosophical thinker; but, until the
relations of the existing with the extinct species and
of the species of the different geographical areas
with one another, were determined with some
exactness, they afforded but an unsafe foundation
for speculation. It was not possible that this
determination should have been effected before

the return of the " Beagle " to England ; and thus
the date which Darwin (writing in 1837) assigns to
the dawn of the new light which was rising in his
mind becomes intelligible.[1]

" In July opened first note-book on Transmutation of Species. Had been greatly struck from
about the month of previous March on character
of South American fossils and species on Galapagos Archipelago. These facts (especially latter)
origin of all my views." (I. p. 276.)

From March, 1837, then, Darwin, not without
many misgivings and fluctuations of opinion,
inclined towards transmutation as a provisional
hypothesis. Three months afterwards he is hard
at work collecting facts for the purpose of testing the hypothesis; and an almost apologetic
passage in a letter to Lyell shows that, already,
the attractions of biology are beginning to predominate over those of geology.

" I have lately been sadly tempted to be idle[2]—

[1] I am indebted to Mr. F. Darwin for the knowledge of a
letter addressed by his father to Dr. Otto Zacharias in 1877
which contains the following paragraph, confirmatory of the
view expressed above : "When I was on board the *Beagle*, I
believed in the permanence of species, but, as far as I can
remember, vague doubts occasionally flitted across my mind.
On my return home in the autumn of 1836, I immediately began
to prepare my journal for publication, and then saw how many
facts indicated the common descent of species, so that in July,
1837, I opened a note-book to record any facts which might bear
on the question. But I did not become convinced that species
were mutable until, I think, two or three years had elapsed."

[2] Darwin generally uses the word "idle" in a peculiar sense.
He means by it working hard at something he likes when he

that is, as far as pure Geology is concerned—by
the delightful number of new views which have
been coming in thickly and steadily—on the
classification and affinities and instincts of animals
—bearing on the question of species. Note-book
after note-book has been filled with facts which
begin to group themselves *clearly* under sub-laws."
(I. p. 298.)

The problem which was to be Darwin's chief
subject of occupation for the rest of his life thus
presented itself, at first, mainly under its distribu-
tional aspect. Why do species present certain relations in space and in time ? Why are the
animals and plants of the Galapagos Archipelago
so like those of South America and yet different
from them ? Why are those of the several islets
more or less different from one another ? Why
are the animals of the latest geological epoch in
South America similar in *facies* to those which
exist in the same region at the present day, and
yet specifically or generically different ?

The reply to these questions, which was almost
universally received fifty years ago, was that animals and plants were created such as they are ;
and that their present distribution, at any rate so
far as terrestrial organisms are concerned, has been
effected by the migration of their ancestors from

ought to be occupied with a less attractive subject. Though it
sounds paradoxical, there is a good deal to be said in favour of
this view of pleasant work.

the region in which the ark stranded after the
subsidence of the deluge. It is true that the
geologists had drawn attention to a good many
tolerably serious difficulties in the way of the
diluvial part of this hypothesis, no less than to the
supposition that the work of creation had occupied
only a brief space of time. But even those, such
as Lyell, who most strenuously argued in favour
of the sufficiency of natural causes for the production of the phenomena of the inorganic world,
held stoutly by the hypothesis of creation in the
case of those of the world of life.

For persons who were unable to feel satisfied
with the fashionable doctrine, there remained only
two alternatives—the hypothesis of spontaneous
generation, and that of descent with modification.
The former was simply the creative hypothesis
with the creator left out; the latter had already
been propounded by De Maillet and Erasmus
Darwin, among others; and, later, systematically
expounded by Lamarck. But in the eyes of the
naturalist of the "Beagle" (and, probably, in those
of most sober thinkers), the advocates of transmutation had done the doctrine they expounded more
harm than good.

Darwin's opinion of the scientific value of the
"Zoonomia" has already been mentioned. His
verdict on Lamarck is given in the following passage of a letter to Lyell (March, 1863):—

"Lastly, you refer repeatedly to my view as a

modification of Lamarck's doctrine of development
and progression. If this is your deliberate opinion
there is nothing to be said, but it does not seem
so to me. Plato, Buffon, my grandfather, before
Lamarck and others, propounded the *obvious* view
that if species were not created separately they
must have descended from other species, and I
can see nothing else in common between the
"Origin" and Lamarck. I believe this way of
putting the case is very injurious to its acceptance,
as it implies necessary progression, and closely
connects Wallace's and my views with what I consider, after two deliberate readings, as a wretched
book, and one from which (I well remember to my
surprise) I gained nothing."

"But," adds Darwin with a little touch of
banter, "I know you rank it higher, which is curious, as it did not in the least shake your belief."
(III. p. 14; see also p. 16, "to me it was an absolutely useless book.")

Unable to find any satisfactory theory of the
process of descent with modification in the works
of his predecessors, Darwin proceeded to lay the
foundations of his own views independently; and
he naturally turned, in the first place, to the only
certainly known examples of descent with modification, namely, those which are presented by
domestic animals and cultivated plants. He
devoted himself to the study of these cases with
a thoroughness to which none of his predecessors

even remotely approximated; and he very soon
had his reward in the discovery "that selection was the keystone of man's success in making useful races of animals and plants." (I. p.
83.)

This was the first step in Darwin's progress,
though its immediate result was to bring him face
to face with a great difficulty. " But how selection
could be applied to organisms living in a state of
nature remained for some time a mystery to me."
(I. p. 83.)

The key to this mystery was furnished by the
accidental perusal of the famous essay of Malthus
" On Population" in the autumn of 1838. The
necessary result of unrestricted multiplication is
competition for the means of existence. The success of one competitor involves the failure of the
rest, that is, their extinction; and this " selection "
is dependent on the better adaptation of the successful competitor to the conditions of the competition. Variation occurs under natural, no less
than under artificial, conditions. Unrestricted
multiplication implies the competition of varieties
and the selection of those which are relatively best
adapted to the conditions.

Neither Erasmus Darwin, nor Lamarck, had any
inkling of the possibility of this process of " natural
selection "; and though it had been foreshadowed
by Wells in 1813, and more fully stated by
Matthew in 1831, the speculations of the latter

writer remained unknown to naturalists until after the publication of the " Origin of Species."

Darwin found in the doctrine of the selection of favourable variations by natural causes, which thus presented itself to his mind, not merely a probable theory of the origin of the diverse species of living forms, but that explanation of the phenomena of adaptation, which previous speculations had utterly failed to give. The process of natural selection is, in fact, dependent on adaptation—it is all one, whether one says that the competitor which survives is the " fittest " or the " best adapted." And it was a perfectly fair deduction that even the most complicated adaptations might result from the summation of a long series of simple favourable variations.

Darwin notes as a serious defect in the first sketch of his theory that he had omitted to consider one very important problem, the solution of which did not occur to him till some time afterwards. " This problem is the tendency in organic beings descended from the same stock to diverge in character as they become modified. . . . The solution, as I believe, is that the modified offspring of all dominant and increasing forms tend to become adapted to many and highly diversified places in the economy of nature." (1. p. 84.)

It is curious that so much importance should be attached to this supplementary idea. It seems obvious that the theory of the origin of species by natural selection necessarily involves the divergence of the forms selected. An individual which varies, *ipso facto* diverges from the type of its species; and its progeny, in which the variation becomes intensified by selection, must diverge still more, not only from the parent stock, but from any other race of that stock starting from a variation of a different character. The selective process could not take place unless the selected variety was either better adapted to the conditions than the original stock, or adapted to other conditions than the original stock. In the first case, the original stock would be sooner or later extirpated; in the second, the type, as represented by the original stock and the variety, would occupy more diversified stations than it did before.

The theory, essentially such as it was published fourteen years later, was written out in 1844, and Darwin was so fully convinced of the importance of his work, as it then stood, that he made special arrangements for its publication in case of his death. But it is a singular example of reticent fortitude, that, although for the next fourteen years the subject never left his mind, and during the latter half of that period he was constantly engaged in amassing facts bearing upon it from wide reading, a colossal correspondence, and a long series of experiments, only two or three friends were cognisant of his views. To the outside world he seemed to have his hands quite sufficiently full of other matters. In 1844, he published his observations on the volcanic islands visited during the voyage of the " Beagle." In 1845, a largely remodelled edition of his " Journal " made its appearance, and immediately won, as it has ever since held, the favour of both the scientific and the unscientific public. In 1846, the " Geological Observations in South America " came out, and this book was no sooner finished than Darwin set to work upon the Cirripedes. He was led to undertake this long and heavy task, partly by his desire to make out the relations of a very anomalous form which he had discovered on the coast of Chili; and partly by a sense of " presumption in accumulating facts and speculating on the subject of variation without having worked out my due share of species." (II. p. 31.) The eight or nine years of labour, which resulted in a monograph of first-rate importance in systematic zoology (to say nothing of such novel points as the discovery of complemental males), left Darwin no room to reproach himself on this score, and few will share his " doubt whether the work was worth the consumption of so much time." (I. p. 82.)

In science no man can safely speculate about the nature and relation of things with which he is unacquainted at first hand, and the acquirement of an intimate and practical knowledge of the process of species-making and of all the uncertainties which underlie the boundaries between species

and varieties, drawn by even the most careful and
conscientious systematists [1] were of no less importance to the author of the "Origin of Species"
than was the bearing of the Cirripede work upon
"the principles of a natural classification." (I. p.
81.) No one, as Darwin justly observes, has a
"right to examine the question of species who
has not minutely described many." (II. p. 39.)

In September, 1854, the Cirripede work was
finished, "ten thousand barnacles" had been sent
"out of the house, all over the world," and Darwin
had the satisfaction of being free to turn again to
his "old notes on species." In 1855, he began to
breed pigeons, and to make observations on the
effects of use and disuse, experiments on seeds,
and so on, while resuming his industrious collection of facts, with a view "to see how far they
favour or are opposed to the notion that wild species
are mutable or immutable. I mean with my
utmost power to give all arguments and facts on
both sides. I have a *number* of people helping
me every way, and giving me most valuable

---

[1] "After describing a set of forms as distinct species, tearing
up my MS., and making them one species, tearing that up and
making them separate, and then making them one again (which
has happened to me), I have gnashed my teeth, cursed species,
and asked what sin I had committed to be so punished." (II.
p. 40.) Is there any naturalist provided with a logical sense and
a large suite of specimens, who has not undergone pangs of the
sort described in this vigorous paragraph, which might, with
advantage, be printed on the title-page of every systematic
monograph as a warning to the uninitiated?

assistance; but I often doubt whether the subject will not quite overpower me." (II. p. 49.)

Early in 1856, on Lyell's advice, Darwin began to write out his views on the origin of species on a scale three or four times as extensive as that of the work published in 1859. In July of the same year he gave a brief sketch of his theory in a letter to Asa Gray; and, in the year 1857, his letters to his correspondents show him to be busily engaged on what he calls his "big book." (II. pp. 85, 94.) In May, 1857, Darwin writes to Wallace: "I am now preparing my work [on the question how and in what way do species and varieties differ from each other] for publication, but I find the subject so very large, that, though I have written many chapters, I do not suppose I shall go to press for two years." (II. p. 95.) In December, 1857, he writes, in the course of a long letter to the same correspondent, "I am extremely glad to hear that you are attending to distribution in accordance with theoretical ideas. I am a firm believer that without speculation there is no good and original observation." (II. p. 108.)[1] In June, 1858, he received from Mr. Wallace, then in the Malay Archipelago, an "Essay on the tendency of varieties to depart indefinitely from

[1] The last remark contains a pregnant truth, but it must be confessed it hardly squares with the declaration in the *Autobiography*, (I. p. 83), that he worked on "true Baconian principles."

the original type," of which Darwin says, "If
Wallace had my MS. sketch written out in 1842
he could not have made a better short abstract!
Even his terms stand now as heads of my chapters.
Please return me the MS., which he does not say
he wishes me to publish, but I shall, of course, at
once write and offer to send it to any journal.
So all my originality, whatever it may amount to,
will be smashed, though my book, if ever it will
have any value, will not be deteriorated; as all
the labour consists in the application of the
theory." (II. p. 116.)

Thus, Darwin's first impulse was to publish
Wallace's essay without note or comment of his
own. But, on consultation with Lyell and Hooker,
the latter of whom had read the sketch of 1844,
they suggested, as an undoubtedly more equitable
course, that extracts from the MS. of 1844 and
from the letter to Dr. Asa Gray should be communicated to the Linnean Society along with
Wallace's essay. The joint communication was
read on July 1, 1858, and published under the
title "On the Tendency of Species to form
Varieties; and on the Perpetuation of Varieties
and Species by Natural Means of Selection."
This was followed, on Darwin's part, by the composition of a summary account of the conclusions
to which his twenty years' work on the species
question had led him. It occupied him for
thirteen months, and appeared in November,

1859, under the title "On the Origin of Species by means of Natural Selection or the Preservation of Favoured Races in the Struggle of Life."

It is doubtful if any single book, except the "Principia," ever worked so great and so rapid a revolution in science, or made so deep an impression on the general mind. It aroused a tempest of opposition and met with equally vehement support, and it must be added that no book has been more widely and persistently misunderstood by both friends and foes. In 1861, Darwin remarks to a correspondent, "You understand my book perfectly, and that I find a very rare event with my critics." (I. p. 313.) The immense popularity which the "Origin" at once acquired was no doubt largely due to its many points of contact with philosophical and theological questions in which every intelligent man feels a profound interest; but a good deal must be assigned to a somewhat delusive simplicity of style, which tends to disguise the complexity and difficulty of the subject, and much to the wealth of information on all sorts of curious problems of natural history, which is made accessible to the most unlearned reader. But long occupation with the work has led the present writer to believe that the "Origin of Species" is one of the hardest of books to master; [1] and he is justified in this

---

[1] He is comforted to find that probably the best qualified judge among all the readers of the *Origin* in 1859 was of the

conviction by observing that although the "Origin" has been close on thirty years before the world, the strangest misconceptions of the essential nature of the theory therein advocated are still put forth by serious writers.

Although, then, the present occasion is not suitable for any detailed criticism of the theory, or of the objections which have been brought against it, it may not be out of place to endeavour to separate the substance of the theory from its accidents; and to show that a variety not only of hostile comments, but of friendly would-be improvements lose their *raison d'être* to the careful student. Observation proves the existence among all living beings of phenomena of three kinds, denoted by the terms heredity, variation, and multiplication. Progeny tend to resemble their parents; nevertheless all their organs and functions are susceptible of departing more or less from the average parental character; and their number is in excess of that of their parents. Severe competition for the means of living, or the struggle for existence, is a necessary consequence of unlimited multiplication; while selection, or the preservation of favourable variations and the extinction of others, is a necessary consequence of severe competition. "Favourable variations" are those which are better adapted to surrounding conditions. It

same opinion. Sir J. Hooker writes, "It is the very hardest book to read, to full profit, that I ever tried." (II. p. 242.)

follows, therefore, that every variety which is selected into a species is so favoured and preserved in consequence of being, in some one or more respects, better adapted to its surroundings than its rivals. In other words, every species which exists, exists in virtue of adaptation, and whatever accounts for that adaptation accounts for the existence of the species.

To say that Darwin has put forward a theory of the adaptation of species, but not of their origin, is therefore to misunderstand the first principles of the theory. For, as has been pointed out, it is a necessary consequence of the theory of selection that every species must have some one or more structural or functional peculiarities, in virtue of the advantage conferred by which, it has fought through the crowd of its competitors and achieved a certain duration. In this sense, it is true that every species has been " originated " by selection.

There is another sense, however, in which it is equally true that selection originates nothing. "Unless profitable variations . . . . occur natural selection can do nothing" (" Origin," Ed. I. p. 82). "Nothing can be effected unless favourable variations occur " (*ibid.*, p. 108). " What applies to one animal will apply throughout time to all animals—that is, if they vary—for otherwise natural selection can do nothing. So it will be with plants " (*ibid.*, p. 113). Strictly speaking, therefore, the origin of species in general lies in variation; while the origin of any particular species lies, firstly, in the occurrence, and secondly, in the selection and preservation of a particular variation. Clearness on this head will relieve one from the necessity of attending to the fallacious assertion that natural selection is a *deus ex machinâ*, or occult agency.

Those, again, who confuse the operation of the natural causes which bring about variation and selection with what they are pleased to call "chance" can hardly have read the opening paragraph of the fifth chapter of the "Origin" (Ed. I, p. 131): "I have sometimes spoken as if the variations . . . . had been due to chance. This is of course a wholly incorrect expression, but it seems to acknowledge plainly our ignorance of the cause of each particular variation."

Another point of great importance to the right comprehension of the theory, is, that while every species must needs have some adaptive advantageous characters to which it owes its preservation by selection, it may possess any number of others which are neither advantageous nor disadvantageous, but indifferent, or even slightly disadvantageous. (*Ibid.*, p. 81.) For variations take place, not merely in one organ or function at a time, but in many; and thus an advantageous variation, which gives rise to the selection of a new race or species, may be accompanied by others which are

indifferent, but which are just as strongly hereditary as the advantageous variations. The advantageous structure is but one product of a modified
general constitution which may manifest itself by
several other products; and the selective process
carries the general constitution along with the
advantageous special peculiarity. A given species
of plant may owe its existence to the selective
adaptation of its flowers to insect fertilisers; but the
character of its leaves may be the result of variations of an indifferent character. It is the origin
of variations of this kind to which Darwin refers in
his frequent reference to what he calls "laws of
correlation of growth " or " correlated variation."

These considerations lead us further to see the
inappropriateness of the objections raised to
Darwin's theory on the ground that natural
selection does not account for the first commencements of useful organs. But it does not pretend
to do so. The source of such commencements is
necessarily to be sought in different variations,
which remain unaffected by selection until they
have taken such a form as to become utilisable in
the struggle for existence.

It is not essential to Darwin's theory that
anything more should be assumed than the facts
of heredity, variation, and unlimited multiplication;
and the validity of the deductive reasoning as to
the effect of the last (that is, of the struggle for
existence which it involves) upon the varieties

Considering the difficulties which surround the question of the causes of variation, it is not to be wondered at, that Darwin should have inclined, sometimes, rather more to one and, sometimes, rather more to another of the possible alternatives. There is little difference between the last edition of the " Origin " (1872) and the first on this head. In 1876, however, he writes to Moritz Wagner, " In my opinion, the greatest error which I have committed has been not allowing sufficient weight to the direct action of the environments, *i.e.*, food, climate, &c., independently of natural selection. . . . . When I wrote the ' Origin,' and for some years afterwards, I could find little good evidence of the direct action of the environment ; now there is a large body of evidence, and your case of the Saturnia is one of the most remarkable of which I have heard." (III, p. 159.) But there is really nothing to prevent the most tenacious adherent to the theory of natural selection from taking any view he pleases as to the importance of the direct influence of conditions and the hereditary transmissibility of the modifications which they produce. In fact, there is a good deal to be said for the view that the so-called direct influence of conditions is itself a case of selection. Whether the hypothesis of Pangenesis be accepted or rejected, it can hardly be doubted that the struggle for existence goes on not merely between distinct organisms, but between the physiological units of which each organism is composed, and that changes in external conditions favour some and hinder others.

After a short stay in Cambridge, Darwin resided in London for the first five years which followed his return to England ; and for three years, he held the post of Secretary to the Geological Society, though he shared to the full his friend Lyell's objection to entanglement in such engagements. In fact, he used to say in later life, more than half in earnest, that he gave up hoping for work from men who accepted official duties and, especially, Government appointments. Happily for him, he was exempted from the necessity of making any sacrifice of this kind, but an even heavier burden was laid upon him. During the earlier half of his voyage Darwin retained the vigorous health of his boyhood, and indeed proved himself to be exceptionally capable of enduring fatigue and privation. An anomalous but severe disorder, which laid him up for several weeks at Valparaiso in 1834, however, seems to have left its mark on his constitution; and, in the later years of his London life, attacks of illness, usually accompanied by severe vomiting and great prostration of strength, became frequent. As he grew older, a considerable part of every day, even at his best times, was spent in misery ; while, not unfrequently, months of suffering rendered work of any kind impossible. Even Darwin's remarkable tenacity of purpose and methodical utilisation of

every particle of available energy could not have
enabled him to achieve a fraction of the vast
amount of labour he got through, in the course of
the following forty years, had not the wisest and the
most loving care unceasingly surrounded him from
the time of his marriage in 1839. As early as
1842, the failure of health was so marked
that removal from London became imperatively
necessary; and Darwin purchased a house and
grounds at Down, a solitary hamlet in Kent, which
was his home for the rest of his life. Under the
strictly regulated conditions of a valetudinarian
existence, the intellectual activity of the invalid
might have put to shame most healthy men; and,
so long as he could hold his head up, there was no
limit to the genial kindness of thought and action
for all about him. Those friends who were
privileged to share the intimate life of the house-
hold at Down have an abiding memory of the
cheerful restfulness which pervaded and characterised it.

After mentioning his settlement at Down,
Darwin writes in his Autobiography :—

"My chief enjoyment and sole employment
throughout life has been scientific work; and the
excitement from such work makes me, for the time,
forget, or drives quite away, my daily discomfort.
I have, therefore, nothing to record during the rest
of my life, except the publication of my several
books." (I, p. 79.)

Of such works published subsequently to 1859, several are monographic discussions of topics briefly dealt with in the " Origin," which, it must always be recollected, was considered by the author to be merely an abstract of an *opus majus.*

The earliest of the books which may be placed in this category, " On the Various Contrivances by which Orchids are Fertilised by Insects," was published in 1862, and whether we regard its theoretical significance, the excellence of the observations and the ingenuity of the reasonings which it records, or the prodigious mass of subsequent investigation of which it has been the parent, it has no superior in point of importance. The conviction that no theory of the origin of species could be satisfactory which failed to offer an explanation of the way in which mechanisms involving adaptations of structure and function to the performance of certain operations are brought about, was, from the first, dominant in Darwin's mind. As has been seen, he rejected Lamarck's views because of their obvious incapacity to furnish such an explanation in the case of the great majority of animal mechanisms, and in that of all those presented by the vegetable world.

So far back as 1793, the wonderful work of Sprengel had established, beyond any reasonable doubt, the fact that, in a large number of cases, a flower is a piece of mechanism the object of which is to convert insect visitors into agents of fertilisa-

must be advantageous in the struggle for existence; and, the more perfect the action of the
mechanism, the greater the advantage. Thus the
way lay open for the operation of natural selection
in gradually perfecting the flower as a fertilisation-
trap. Analogous reasoning applies to the fertil-
ising insect. The better its structure is adapted
to that of the trap, the more will it be able to
profit by the bait, whether of honey or of pollen,
to the exclusion of its competitors. Thus, by a
sort of action and reaction, a two-fold series of
adaptive modifications will be brought about.

In 1865, the important bearing of this subject
on his theory led Darwin to commence a great
series of laborious and difficult experiments on the
fertilisation of plants, which occupied him for
eleven years, and furnished him with the unex-
pectedly strong evidence in favour of the influence
of crossing which he published in 1876, under the
title of "The Effects of Cross and Self Fertilisation
in the Vegetable Kingdom." Incidentally, as it
were, to this heavy piece of work, he made the
remarkable series of observations on the different
arrangements by which crossing is favoured and,
in many cases, necessitated, which appeared in the
work on "The Different Forms of Flowers in
Plants of the same Species" in 1877.

In the course of the twenty years during which
Darwin was thus occupied in opening up new
regions of investigation to the botanist and

showing the profound physiological significance of
the apparently meaningless diversities of floral
structure, his attention was keenly alive to any
other interesting phenomena of plant life which
came in his way. In his correspondence, he not
unfrequently laughs at himself for his ignorance
of systematic botany; and his acquaintance with
vegetable anatomy and physiology was of the
slenderest. Nevertheless, if any of the less
common features of plant life came under his
notice, that imperious necessity of seeking for
causes which nature had laid upon him, impelled,
and indeed compelled, him to inquire the how
and the why of the fact, and its bearing on his
general views. And as, happily, the atavic tendency to frame hypotheses was accompanied
by an equally strong need to test them by well-
devised experiments, and to acquire all possible
information before publishing his results, the
effect was that he touched no topic without
elucidating it.

Thus the investigation of the operations of
insectivorous plants, embodied in the work on that
topic published in 1875, was started fifteen years
before, by a passing observation made during one
of Darwin's rare holidays.

"In the summer of 1860, I was idling and
resting near Hartfield, where two species of
Drosera abound; and I noticed that numerous
insects had been entrapped by the leaves. I

carried home some plants, and on giving them some insects saw the movements of the tentacles, and this made me think it possible that the insects were caught for some special purpose. Fortunately, a crucial test occurred to me, that of placing a large number of leaves in various nitrogenous and non-nitrogenous fluids of equal density; and as soon as I found that the former alone excited energetic movements, it was obvious that here was a fine new field for investigation." (I, p. 95.)

The researches thus initiated led to the proof that plants are capable of secreting a digestive fluid like that of animals, and of profiting by the result of digestion; whereby the peculiar apparatuses of the insectivorous plants were brought within the scope of natural selection. Moreover, these inquiries widely enlarged our knowledge of the manner in which stimuli are transmitted in plants, and opened up a prospect of drawing closer the analogies between the motor processes of plants and those of animals.

So with respect to the books on "Climbing Plants" (1875), and on the "Power of Movement in Plants" (1880), Darwin says;—

"I was led to take up this subject by reading a short paper by Asa Gray, published in 1858. He sent me some seeds, and on raising some plants I was so much fascinated and perplexed by the revolving movements of the tendrils and stems, which movements are really very simple, though

appearing at first sight very complex, that I pro-
cured various other kinds of climbing plants
and studied the whole subject. . . . Some of the
adaptations displayed by climbing plants are as
beautiful as those of orchids for ensuring cross-
fertilisation." (I, p. 93.)

In the midst of all this amount of work,
remarkable alike for its variety and its importance,
among plants, the animal kingdom was by no
means neglected. A large moiety of "The
Variation of Animals and Plants under Domestication" (1868), which contains the *pièces justifica-
tives* of the first chapter of the " Origin," is devoted
to domestic animals, and the hypothesis of
"pangenesis" propounded in the second volume
applies to the whole living world. In the "Origin" Darwin throws out some suggestions as to
the causes of variation, but he takes heredity, as it
is manifested by individual organisms, for granted,
as an ultimate fact; pangenesis is an attempt to
account for the phenomena of heredity in the
organism, on the assumption that the physiological
units of which the organism is composed give off
gemmules, which, in virtue of heredity, tend to
reproduce the unit from which they are derived.

That Darwin had the application of his theory
to the origin of the human species clearly in his
mind in 1859, is obvious from a passage in the
first edition of "The Origin of Species." (Ed. I,
p. 488.) "In the distant future I see open fields

for far more important researches. Psychology
will be based on a new foundation, that of the
necessary acquirement of each mental power and
capacity by gradation. Light will be thrown on
the origin of man and his history." It is one of
the curiosities of scientific literature, that, in the
face of this plain declaration, its author should have
been charged with concealing his opinions on the
subject of the origin of man. But he reserved the
full statement of his views until 1871, when the
"Descent of Man" was published. The "Expression of the Emotions" (originally intended to form
only a chapter in the "Descent of Man") grew into
a separate volume, which appeared in 1872.
Although always taking a keen interest in geology,
Darwin naturally found no time disposable for
geological work, even had his health permitted it,
after he became seriously engaged with the great
problem of species. But the last of his labours is,
in some sense, a return to his earliest, inasmuch as
it is an expansion of a short paper read before the
Geological Society more than forty years before,
and, as he says, "revived old geological thoughts"
(I, p. 98). In fact, "The Formation of Vegetable
Mould through the Action of Worms," affords as
striking an example of the great results produced
by the long-continued operation of small causes as
even the author of the "Principles of Geology"
could have desired.

In the early months of 1882 Darwin's health

attention of all and the curiosity of all have been probably more or less excited on the subject of that work. All I can do, and all I shall attempt to do, is to put before you that kind of judgment which has been formed by a man, who, of course, is liable to judge erroneously ; but, at any rate, of one whose business and profession it is to form judgments upon questions of this nature.

And here, as it will always happen when dealing with an extensive subject, the greater part of my course—if, indeed, so small a number of lectures can be properly called a course—must be devoted to preliminary matters, or rather to a statement of those facts and of those principles which the work itself dwells upon, and brings more or less directly before us. I have no right to suppose that all or any of you are naturalists ; and, even if you were, the misconceptions and misunderstandings prevalent even among naturalists, on these matters, would make it desirable that I should take the course I now propose to take,—that I should start from the beginning,—that I should endeavour to point out what is the existing state of the organic world—that I should point out its past condition,—that I should state what is the precise nature of the undertaking which Mr. Darwin has taken in hand ; that I should endeavour to show you what are the only methods by which that undertaking can be brought to an issue, and to point out to you how far the author of the work

in question has satisfied those conditions, how far
he has not satisfied them, how far they are satisfiable by man, and how far they are not satisfiable
by man.

To-night, in taking up the first part of the
question, I shall endeavour to put before you a
sort of broad notion of our knowledge of the condition of the living world. There are many ways
of doing this. I might deal with it pictorially and
graphically. Following the example of Humboldt
in his " Aspects of Nature," I might endeavour to
point out the infinite variety of organic life in
every mode of its existence, with reference to the
variations of climate and the like ; and such an
attempt would be fraught with interest to us all ;
but considering the subject before us, such a course
would not be that best calculated to assist us. In
an argument of this kind we must go further and
dig deeper into the matter ; we must endeavour to
look into the foundations of living Nature, if I
may so say, and discover the principles involved in
some of her most secret operations. I propose,
therefore, in the first place, to take some ordinary
animal with which you are all familiar, and, by
easily comprehensible and obvious examples drawn
from it, to show what are the kind of problems
which living beings in general lay before us ; and
I shall then show you that the same problems are
laid open to us by all kinds of living beings.
But, first, let me say in what sense I have used the

words " organic nature." In speaking of the
causes which lead to our present knowledge of
organic nature, I have used it almost as an
equivalent of the word " living," and for this
reason,—that in almost all living beings you can
distinguish several distinct portions set apart to
do particular things and work in a particular way.
These are termed " organs," and the whole
together is called " organic." And as it is
universally characteristic of them, the term
" organic " has been very conveniently employed
to denote the whole of living nature,—the whole
of the plant world, and the whole of the animal
world.

Few animals can be more familiar to you than
that whose skeleton is shown on our diagram.
You need not bother yourselves with this " *Equus
caballus*" written under it ; that is only the Latin
name of it, and does not make it any better. It
simply means the common horse. Suppose we
wish to understand all about the horse. Our
first object must be to study the structure of the
animal. The whole of his body is inclosed within
a hide, a skin covered with hair ; and if that hide
or skin be taken off, we find a great mass of flesh,
or what is technically called muscle, being the
substance which by its power of contraction enables
the animal to move. These muscles move the hard
parts one upon the other, and so give that strength
and power of motion which renders the horse so

useful to us in the performance of those services in which we employ him.

And then, on separating and removing the whole of this skin and flesh, you have a great series of bones, hard structures, bound together with ligaments, and forming the skeleton which is represented here.

In that skeleton there are a number of parts to be recognised. The long series of bones, beginning from the skull and ending in the tail, is called the spine, and those in front are the ribs; and then there are two pairs of limbs, one before and one behind; and there are what we all know as the fore-legs and the hind-legs. If we pursue our researches into the interior of this animal, we find within the framework of the skeleton a great cavity, or rather, I should say, two great cavities, —one cavity beginning in the skull and running through the neck-bones, along the spine, and ending in the tail, containing the brain and the spinal marrow, which are extremely important organs. The second great cavity, commencing with the mouth, contains the gullet, the stomach, the long intestine, and all the rest of those internal apparatus which are essential for digestion; and then in the same great cavity, there are lodged the heart and all the great vessels going from it; and, besides that, the organs of respiration—the lungs: and then the kidneys, and the organs of reproduction, and so on. Let us now endeavour to reduce this notion of a horse that we now have, to some such kind of simple expressions as can be at once, and without difficulty, retained in the mind, apart from all minor details.   If I make a transverse section, that is, if I were to saw a dead horse across, I should find that, if I left out the details, and supposing I took my section through the anterior region, and through the fore-limbs, I should have here this kind of section of the body (Fig. 1).   Here would be the upper part of the animal—that great mass of bones that we spoke of as the spine (*a*, Fig. 1). Here I should have the alimentary canal (*b*, Fig. 1). Here I should have the heart (*c*, Fig. 1); and then you see, there would be a kind of double tube, the whole being inclosed with-in the hide; the spinal marrow would be placed in the upper tube (*a*, Fig. 1), and in the lower tube (*d d*, Fig. 1), there would be the alimentary canal (*b*), and the heart (*c*); and here I shall have the legs proceeding from each side.   For simplicity's sake, I represent them merely as

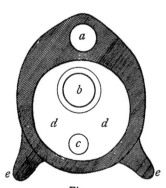

Fig. 1.

stumps (*e e*, Fig. 1). Now that is a horse—as
mathematicians would say—reduced to its most
simple expression. Carry that in your minds, if
you please, as a simplified idea of the structure of
the horse. The considerations which I have now
put before you belong to what we technically call
the "Anatomy" of the horse. Now, suppose we
go to work upon these several parts,—flesh and
hair, and skin and bone, and lay open these various
organs with our scalpels, and examine them by
means of our magnifying-glasses, and see what we
can make of them. We shall find that the flesh
is made up of bundles of strong fibres. The brain
and nerves, too, we shall find, are made up of
fibres, and these queer-looking things that are
called ganglionic corpuscles. If we take a slice of
the bone and examine it, we shall find that it is
very like this diagram of a section of the bone of
on ostrich, though differing, of course, in some
details ; and if we take any part whatsoever of the
tissue, and examine it, we shall find it all has a
minute structure, visible only under the microscope.
All these parts constitute microscopic anatomy or
"Histology." These parts are constantly being
changed ; every part is constantly growing, decaying, and being replaced during the life of the animal.
The tissue is constantly replaced by new material ;
and if you go back to the young state of the tissue
in the case of muscle, or in the case of skin, or any
of the organs I have mentioned, you will find that

term technically its Morphology), I must now turn
to another aspect. A horse is not a mere dead
structure : it is an active, living, working machine.
Hitherto we have, as it were, been looking at a
steam-engine with the fires out, and nothing in the
boiler; but the body of the living animal is a
beautifully-formed active machine, and every part
has its different work to do in the working of that
machine, which is what we call its life. The
horse, if you see him after his day's work is done,
is cropping the grass in the fields, as it may be, or
munching the oats in his stable. What is he
doing? His jaws are working as a mill—and a
very complex mill too—grinding the corn, or
crushing the grass to a pulp. As soon as that
operation has taken place, the food is passed down
to the stomach, and there it is mixed with the
chemical fluid called the gastric juice, a substance
which has the peculiar property of making soluble
and dissolving out the nutritious matter in the
grass, and leaving behind those parts which are
not nutritious; so that you have, first, the mill,
then a sort of chemical digester; and then the
food, thus partially dissolved, is carried back
by the muscular contractions of the intestines into
the hinder parts of the body, while the soluble
portions are taken up into the blood. The blood
is contained in a vast system of pipes, spreading
through the whole body, connected with a force-
pump,—the heart,—which, by its position and by

backbone; and to this spinal cord are attached a number of fibres termed nerves, which proceed to all parts of the structure. By means of these the eyes, nose, tongue, and skin—all the organs of perception—transmit impressions or sensations to the brain, which acts as a sort of great central telegraph-office, receiving impressions and sending messages to all parts of the body, and putting in motion the muscles necessary to accomplish any movement that may be desired. So that you have here an extremely complex and beautifully-proportioned machine, with all its parts working harmoniously together towards one common object—the preservation of the life of the animal.

Now, note this: the horse makes up its waste by feeding, and its food is grass or oats, or perhaps other vegetable products; therefore, in the long run, the source of all this complex machinery lies in the vegetable kingdom. But where does the grass, or the oat, or any other plant, obtain this nourishing food-producing material? At first it is a little seed, which soon begins to draw into itself from the earth and the surrounding air matters which in themselves contain no vital properties whatever; it absorbs into its own substance water, an inorganic body; it draws into its substance carbonic acid, an inorganic matter; and ammonia, another inorganic matter, found in the air; and then, by some wonderful chemical process, the details of which chemists do not yet understand, though they are near foreshadowing them, it combines them into one substance, which is known to us as " Protein," a complex compound of carbon, hydrogen, oxygen, and nitrogen, which alone possesses the property of manifesting vitality and of permanently supporting animal life. So that, you see, the waste products of the animal economy, the effete materials which are continually being thrown off by all living beings, in the form of organic matters, are constantly replaced by supplies of the necessary repairing and rebuilding materials drawn from the plants, which in their turn manufacture them, so to speak, by a mysterious .combination of those same inorganic materials.

Let us trace out the history of the horse in another direction. After a certain time, as the result of sickness or disease, the effect of accident, or the consequence of old age, sooner or later, the animal dies. The multitudinous operations of this beautiful mechanism flag in their performance, the horse loses its vigour, and after passing through the curious series of changes comprised in its formation and preservation, it finally decays, and ends its life by going back into that inorganic world from which all but an inappreciable fraction of its substance was derived. Its bones become mere carbonate and phosphate of lime ; the matter of its flesh, and of its other parts, becomes, in the long run, converted into carbonic acid, into water, and into ammonia. You will now, perhaps, understand the curious relation of the animal with the plant, of the organic with the inorganic world, which is shown in this diagram.

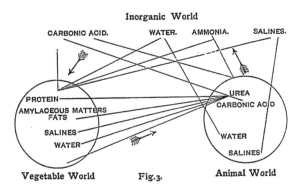

Fig. 3.

The plant gathers these inorganic materials together and makes them up into its own substance. The animal eats the plant and appropriates the nutritious portions to its own sustenance, rejects and gets rid of the useless matters; and, finally, the animal itself dies, and its whole body is decomposed and returned into the inorganic world. There is thus a constant circulation from one to the other, a continual formation of organic life from inorganic matters, and as constant a return of the matter of living bodies to the inorganic world; so that the materials of which our bodies are composed are largely, in all probability, the substances which constituted the matter of long extinct creations, but which have in the interval constituted a part of the inorganic world.

Thus we come to the conclusion, strange at first sight, that the MATTER constituting the living world is identical with that which forms the inorganic world. And not less true is it that, remarkable as are the powers or, in other words, as are the FORCES which are exerted by living beings, yet all these forces are either identical with those which exist in the inorganic world, or they are convertible into them; I mean in just the same sense as the researches of physical philosophers have shown that heat is convertible into electricity, that electricity is convertible into magnetism, magnetism into mechanical force or chemical force, and any one of them with the other, each being measurable in terms of the other, —even so, I say, that great law is applicable to the living world. Consider why is the skeleton of this horse capable of supporting the masses of flesh and the various organs forming the living body, unless it is because of the action of the same forces of cohesion which combines together the particles of matter composing this piece of chalk? What is there in the muscular contractile power of the animal but the force which is expressible, and which is in a certain sense convertible, into

the force of gravity which it overcomes ? Or, if
you go to more hidden processes, in what does the
process of digestion differ from those processes
which are carried on in the laboratory of the
chemist ? Even if we take the most recondite
and most complex operations of animal life—those
of the nervous system, these of late years have
been shown to be—I do not say identical in any
sense with the electrical processes—but this has
been shown, that they are in some way or other
associated with them; that is to say, that every
amount of nervous action is accompanied by a
certain amount of electrical disturbance in the
particles of the nerves in which that nervous
action is carried on. In this way the nervous
action is related to electricity in the same way
that heat is related to electricity ; and the same
sort of argument which demonstrates the two latter
to be related to one another shows that the nervous
forces are correlated to electricity ; for the experiments of M. Dubois Reymond and others have
shown that whenever a nerve is in a state of
excitement, sending a message to the muscles or
conveying an impression to the brain, there is a
disturbance of the electrical condition of that
nerve which does not exist at other times ; and
there are a number of other facts and phenomena
of that sort; so that we come to the broad conclusion that not only as to living matter itself, but
as to the forces that matter exerts, there is a close

grain in weight, undergoes a series of changes,—
wonderful, complex changes. Finally, upon its
surface there is fashioned a little elevation, which
afterwards becomes divided and marked by a
groove. The lateral boundaries of the groove
extend upwards and downwards, and at length
give rise to a double tube. In the upper and
smaller tube the spinal marrow and brain are
fashioned ; in the lower, the alimentary canal and
heart ; and at length two pairs of buds shoot out at
the sides of the body, and they are the rudiments
of the limbs. In fact a true drawing of a section
of the embryo in this state would in all essential
respects resemble that diagram of a horse reduced
to its simplest expression, which I first placed
before you (Fig. 1).

Slowly and gradually these changes take place.
The whole of the body, at first, can be broken up
into " cells," which become in one place metamorphosed into muscle,—in another place into
gristle and bone,—in another place into fibrous
tissue,—and in another into hair ; every part
becoming gradually and slowly fashioned, as if
there were an artificer at work in each of these
complex structures that I have mentioned. This
embryo, as it is called, then passes into other conditions. I should tell you that there is a time when
the embryos of neither dog, nor horse, nor porpoise,
nor monkey, nor man, can be distinguished by any
essential feature one from the other ; there is a

time when they each and all of them resemble this one of the dog. But as development advances, all the parts acquire their speciality, till at length you have the embryo converted into the form of the parent from which it started. So that you see, this living animal, this horse, begins its existence as a minute particle of nitrogenous matter, which, being supplied with nutriment (derived, as I have shown, from the inorganic world), grows up according to the special type and construction of its parents, works and undergoes a constant waste, and that waste is made good by nutriment derived from the inorganic world; the waste given off in this way being directly added to the inorganic world. Eventually the animal itself dies, and, by the process of decomposition, its whole body is returned to those conditions of inorganic matter in which its substance originated.

This, then, is that which is true of every living form, from the lowest plant to the highest animal —to man himself. You might define the life of every one in exactly the same terms as those which I have now used; the difference between the highest and the lowest being simply in the complexity of the developmental changes, the variety of the structural forms, and the diversity of the physiological functions which are exerted by each.

If I were to take an oak tree, as a specimen of

the plant world, I should find that it originated in an acorn, which, too, commenced in a cell; the acorn is placed in the ground, and it very speedily begins to absorb the inorganic matters I have named, adds enormously to its bulk, and we can see it, year after year, extending itself upward and downward, attracting and appropriating to itself inorganic materials, which it vivifies, and eventually, as it ripens, gives off its own proper acorns, which again run the same course. But I need not multiply examples,—from the highest to the lowest the essential features of life are the same as I have described in each of these cases.

So much, then, for these particular features of the organic world, which you can understand and comprehend, so long as you confine yourself to one sort of living being, and study that only.

But, as you know, horses are not the only living creatures in the world; and again, horses, like all other animals, have certain limits—are confined to a certain area on the surface of the earth on which we live,—and, as that is the simpler matter, I may take that first. In its wild state, and before the discovery of America, when the natural state of things was interfered with by the Spaniards, the horse was only to be found in parts of the earth which are known to geographers as the Old World; that is to say, you might meet with horses in Europe, Asia, or Africa; but there were none in Australia, and there were none whatsoever

in the whole continent of America, from Labrador
down to Cape Horn. This is an empirical fact, and
it is what is called, stated in the way I have
given it you, the " Geographical Distribution " of
the horse.

Why horses should be found in Europe, Asia,
and Africa, and not in America, is not obvious;
the explanation that the conditions of life in
America are unfavourable to their existence, and
that, therefore, they had not been created there,
evidently does not apply ; for when the invading
Spaniards, or our own yeomen farmers, conveyed
horses to these countries for their own use, they
were found to thrive well and multiply very
rapidly ; and many are even now running wild in
those countries, and in a perfectly natural condition.
Now, suppose we were to do for every animal
what we have here done for the horse,—that is,
to mark off and distinguish the particular district
or region to which each belonged ; and supposing
we tabulated all these results, that would be
called the Geographical Distribution of animals,
while a corresponding study of plants would yield
as a result the Geographical Distribution of
plants.

I pass on from that now, as I merely wished to
explain to you what I meant by the use of the
term " Geographical Distribution." As I said,
there is another aspect, and a much more important one, and that is, the relations of the various

animals to one another. The horse is a very well-
defined matter-of-fact sort of animal, and we are
all pretty familiar with its structure. I dare say
it may have struck you, that it resembles very
much no other member of the animal kingdom,
except perhaps the zebra or the ass. But let me
ask you to look along these diagrams. Here is
the skeleton of the horse, and here the skeleton
of the dog. You will notice that we have in the
horse a skull, a backbone and ribs, shoulder-blades
and haunch-bones. In the fore-limb, one upper
arm-bone, two fore arm-bones, wrist-bones (wrongly
called knee), and middle hand-bones, ending in
the three bones of a finger, the last of which is
sheathed in the horny hoof of the fore-foot : in the
hind-limb, one thigh-bone, two leg-bones, ankle-
bones, and middle foot-bones, ending in the three
bones of a toe, the last of which is encased in the
hoof of the hind-foot. Now turn to the dog's
skeleton. We find identically the same bones, but
more of them, there being more toes in each foot,
and hence more toe-bones.

Well, that is a very curious thing ! The fact is
that the dog and the horse—when one gets a
look at them without the outward impediments of
the skin—are found to be made in very much the
same sort of fashion. And if I were to make a
transverse section of the dog, I should find the
same organs that I have already shown you as
forming parts of the horse. Well, here is another

skeleton—that of a kind of lemur—you see he
has just the same bones; and if I were to make a
transverse section of it, it would be just the same
again. In your mind's eye turn him round, so as
to put his backbone in a position inclined obliquely
upwards and forwards, just as in the next three
diagrams, which represent the skeletons of an
orang, a chimpanzee, and a gorilla, and you find
you have no trouble in identifying the bones
throughout; and lastly turn to the end of the
series, the diagram representing a man's skeleton,
and still you find no great structural feature
essentially altered. There are the same bones in
the same relations. From the horse we pass on
and on, with gradual steps until we arrive at last
at the highest known forms. On the other hand,
take the other line of diagrams, and pass from the
horse downwards in the scale to this fish; and
still, though the modifications are vastly greater,
the essential framework of the organisation
remains unchanged. Here, for instance, is a
porpoise: here is its strong backbone, with the
cavity running through it, which contains the
spinal cord; here are the ribs, here the shoulder-
blade; here is the little short upper-arm bone,
here are the two forearm bones, the wrist-bone,
and the finger-bones.

Strange, is it not, that the porpoise should have
in this queer-looking affair—its flapper (as it is
called), the same fundamental elements as the

fore-leg of the horse or the dog, or the ape or
man; and here you will notice a very curious
thing,—the hinder limbs are absent. Now, let
us make another jump. Let us go to the codfish:
here you see is the forearm, in this large pectoral fin
—carrying your mind's eye onward from the flapper
of the porpoise. And here you have the hinder
limbs restored in the shape of these ventral fins.
If I were to make a transverse section of this, I
should find just the same organs that we have
before noticed. So that, you see, there comes out
this strange conclusion as the result of our
investigations, that the horse, when examined
and compared with other animals, is found by no
means to stand alone in Nature; but that there
are an enormous number of other creatures which
have backbones, ribs, and legs, and other parts
arranged in the same general manner, and in
all their formation exhibiting the same broad
peculiarities.

I am sure that you cannot have followed me
even in this extremely elementary exposition of
the structural relations of animals, without seeing
what I have been driving at all through, which is,
to show you that, step by step, naturalists have
come to the idea of a unity of plan, or conformity
of construction, among animals which appeared at
first sight to be extremely dissimilar.

And here you have evidence of such a unity of
plan among all the animals which have backbones,

and which we technically call *Vertebrata*. But there are multitudes of other animals, such as crabs, lobsters, spiders, and so on, which we term *Annulosa*. In these I could not point out to you the parts that correspond with those of the horse,—the backbone, for instance,—as they are constructed upon a very different principle, which is also common to all of them; that is to say, the lobster, the spider, and the centipede, have a common plan running through their whole arrangement, in just the same way that the horse, the dog, and the porpoise assimilate to each other.

Yet other creatures—whelks, cuttlefishes, oysters, snails, and all their tribe (*Mollusca*)—resemble one another in the same way, but differ from both *Vertebrata* and *Annulosa*; and the like is true of the animals called *Cœlenterata* (Polypes) and *Protozoa* (animalcules and sponges).

Now, by pursuing this sort of comparison, naturalists have arrived at the conviction that there are,—some think five, and some seven,—but certainly not more than the latter number—and perhaps it is simpler to assume five—distinct plans or constructions in the whole of the animal world; and that the hundreds of thousands of species of creatures on the surface of the earth, are all reducible to those five, or, at most, seven, plans of organisation.

But can we go no further than that? When one has got so far, one is tempted to go on a step

and inquire whether we cannot go back yet
further and bring down the whole to modifications
of one primordial unit. The anatomist cannot do
this ; but if he call to his aid the study of development, he can do it. For we shall find that, distinct as those plans are, whether it be a porpoise
or man, or lobster, or any of those other kinds I
have mentioned, every one begins its existence
with one and the same primitive form,—that of
the egg, consisting, as we have seen, of a nitrogenous substance, having a small particle or nucleus
in the centre of it. Furthermore, the earlier
changes of each are substantially the same. And
it is in this that lies that true "unity of organisation" of the animal kingdom which has been
guessed at and fancied for many years ; but which
it has been left to the present time to be demonstrated by the careful study of development. But
is it possible to go another step further still, and
to show that in the same way the whole of the
organic world is reducible to one primitive condition of form ? Is there among the plants the
same primitive form of organisation, and is that
identical with that of the animal kingdom ? The
reply to that question, too, is not uncertain or
doubtful. It is now proved that every plant
begins its existence under the same form ; that is
to say, in that of a cell—a particle of nitrogenous
matter having substantially the same conditions.
So that if you trace back the oak to its first

germ, or a man, or a horse, or lobster, or oyster, or
any other animal you choose to name, you shall find
each and all of these commencing their existence
in forms essentially similar to each other; and,
furthermore, that the first processes of growth,
and many of the subsequent modifications, are
essentially the same in principle in almost all.

In conclusion, let me, in a few words, recapitu-
late the positions which I have laid down.   And
you must understand that I have not been
talking mere theory; I have been speaking of
matters which are as plainly demonstrable as the
commonest propositions of Euclid—of facts that
must form the basis of all speculations and beliefs
in Biological science.   We have gradually traced
down all organic forms, or, in other words, we have
analysed the present condition of animated nature,
until we found that each species took its origin in
a form similar to that under which all the others
commenced their existence.   We have found the
whole of the vast array of living forms with which
we are surrounded, constantly growing, increasing,
decaying and disappearing; the animal constantly
attracting, modifying, and applying to its sustenance the matter of the vegetable kingdom, which
derived its support from the absorption and conversion of inorganic matter.   And so constant and
universal is this absorption, waste, and reproduction, that it may be said with perfect certainty
that there is left in no one of our bodies at the

present moment a millionth part of the matter of which they were originally formed! We have seen, again, that not only is the living matter derived from the inorganic world, but that the forces of that matter are all of them correlative with and convertible into those of inorganic nature.

This, for our present purposes, is the best view of the present condition of organic nature which I can lay before you : it gives you the great outlines of a vast picture, which you must fill up by your own study.

In the next lecture I shall endeavour in the same way to go back into the past, and to sketch in the same broad manner the history of life in epochs preceding our own.

II

THE PAST CONDITION OF ORGANIC NATURE.

IN the lecture which I delivered last Monday
evening, I endeavoured to sketch in a very brief
manner, but as well as the time at my disposal
would permit, the present condition of organic
nature, meaning by that large title simply an
indication of the great, broad, and general
principles which are to be discovered by those
who look attentively at the phenomena of organic
nature as at present displayed. The general
result of our investigations might be summed up
thus : we found that the multiplicity of the forms
of animal life, great as that may be, may be
reduced to a comparatively few primitive plans or
types of construction ; that a further study of the
development of those different forms revealed to
us that they were again reducible, until we at
last brought the infinite diversity of animal, and
even vegetable life, down to the primordial form
of a single cell.

We found that our analysis of the organic
world, whether animals or plants, showed, in the
long run, that they might both be reduced into,
and were, in fact, composed of, the same constituents. And we saw that the plant obtained
the materials constituting its substance by a
peculiar combination of matters belonging entirely
to the inorganic world; that, then, the animal was
constantly appropriating the nitrogenous matters
of the plant to its own nourishment, and returning
them back to the inorganic world, in what we
spoke of as its waste; and that finally, when the
animal ceased to exist, the constituents of its body
were dissolved and transmitted to that inorganic
world whence they had been at first abstracted.
Thus we saw in both the blade of grass and the
horse but the same elements differently combined
and arranged. We discovered a continual circulation going on,—the plant drawing in the elements
of inorganic nature and combining them into food
for the animal creation; the animal borrowing
from the plant the matter for its own support,
giving off during its life products which returned
immediately to the inorganic world; and that,
eventually, the constituent materials of the whole
structure of both animals and plants were thus
returned to their original source : there was a
constant passage from one state of existence to
another, and a returning back again.

Lastly, when we endeavoured to form some

notion of the nature of the forces exercised by
living beings, we discovered that they—if not
capable of being subjected to the same minute
analysis as the constituents of those beings themselves—that they were correlative with—that they
were the equivalents of the forces of inorganic
nature—that they were, in the sense in which the
term is now used, convertible with them. That was
our general result.

And now, leaving the Present, I must endeavour
in the same manner to put before you the facts
that are to be discovered in the Past history of
the living world, in the past conditions of organic
nature.   We have, to-night, to deal with the facts
of that history—a history involving periods of
time before which our mere human records sink
into utter insignificance—a history the variety and
physical magnitude of whose events cannot even
be foreshadowed by the history of human life and
human phenomena—a history of the most varied
and complex character.

We must deal with the history, then, in the
first place, as we should deal with all other
histories.   The historical student knows that his
first business should be to inquire into the validity
of his evidence, and the nature of the record in
which the evidence is contained, that he may be
able to form a proper estimate of the correctness
of the conclusions which have been drawn from
that evidence.   So, here, we must pass, in the first

place, to the consideration of a matter which may seem foreign to the question under discussion. We must dwell upon the nature of the records, and the credibility of the evidence they contain; we must look to the completeness or incompleteness of those records themselves, before we turn to that which they contain and reveal. The question of the credibility of the history, happily for us, will not require much consideration, for, in this history, unlike those of human origin, there can be no cavilling, no differences as to the reality and truth of the facts of which it is made up; the facts state themselves, and are laid out clearly before us.

But, although one of the greatest difficulties of the historical student is cleared out of our path. there are other difficulties—difficulties in rightly interpreting the facts as they are presented to us —which may be compared with the greatest difficulties of any other kinds of historical study.

What is this record of the past history of the globe, and what are the questions which are involved in an inquiry into its completeness or incompleteness? That record is composed of mud; and the question which we have to investigate this evening resolves itself into a question of the formation of mud. You may think, perhaps, that this is a vast step—of almost from the sublime to the ridiculous—from the contemplation of the history of the past ages of the world's existence to the consideration of the history of the formation of mud! But, in Nature, there is nothing mean and unworthy of attention; there is nothing ridiculous or contemptible in any of her works; and this inquiry, you will soon see, I hope, takes us to the very root and foundations of our subject.

How, then, is mud formed? Always, with some trifling exceptions, which I need not consider now—always, as the result of the action of water, wearing down and disintegrating the surface of the earth and rocks with which it comes in contact—pounding and grinding it down, and carrying the particles away to places where they cease to be disturbed by this mechanical action, and where they can subside and rest. For the ocean, urged by winds, washes, as we know, a long extent of coast, and every wave, loaded as it is with particles of sand and gravel as it breaks upon the shore, does something towards the disintegrating process. And thus, slowly but surely, the hardest rocks are gradually ground down to a powdery substance; and the mud thus formed, coarser or finer, as the case may be, is carried by the rush of the tides, or currents, till it reaches the comparatively deeper parts of the ocean, in which it can sink to the bottom, that is, to parts where there is a depth of about fourteen or fifteen fathoms, a depth at which the water is, usually, nearly motionless, and in which, of course, the finer particles of this detritus, or mud as we call it, sinks to the bottom.

Or, again, if you take a river, rushing down from its mountain sources, brawling over the stones and rocks that intersect its path, loosening, removing, and carrying with it in its downward course the pebbles and lighter matters from its banks, it crushes and pounds down the rocks and earths in precisely the same way as the wearing action of the sea waves. The matters forming the deposit are torn from the mountain-side and whirled impetuously into the valley, more slowly over the plain, thence into the estuary, and from the estuary they are swept into the sea. The coarser and heavier fragments are obviously deposited first, that is, as soon as the current begins to lose its force by becoming amalgamated with the stiller depths of the ocean, but the finer and lighter particles are carried further on, and eventually deposited in a deeper and stiller portion of the ocean.

It clearly follows from this that mud gives us a chronology; for it is evident that supposing this, which I now sketch, to be the sea bottom, and supposing this to be a coast-line ; from the washing action of the sea upon the rock, wearing and grinding it down into a sediment of mud, the mud will be carried down, and, at length, deposited in the deeper parts of this sea bottom, where it will form a layer ; and then, while that first layer is hardening, other mud which is coming from the same source will, of course, be carried to the same place; and, as it is quite impossible for it to get beneath the layer already there, it deposits itself above it, and forms another layer, and in that way you gradually have layers of mud constantly forming and hardening one above the other, and conveying a record of time.

It is a necessary result of the operation of the law of gravitation that the uppermost layer shall be the youngest and the lowest the oldest, and that the different beds shall be older at any particular point or spot in exactly the ratio of their depth from the surface. So that if they were upheaved afterwards, and you had a series of these different layers of mud, converted into sandstone, or limestone, as the case might be, you might be sure that the bottom layer was deposited first, and that the upper layers were formed afterwards. Here, you see, is the first step in the history ——these layers of mud give us an idea of time.

The whole surface of the earth,—I speak broadly, and leave out minor qualifications,—is made up of such layers of mud, so hard, the majority of them, that we call them rock whether limestone or sandstone, or other varieties of rock. And, seeing that every part of the crust of the earth is made up in this way, you might think that the determination of the chronology, the fixing of the time which it has taken to form this crust is a comparatively simple matter. Take a broad average, ascertain how fast the mud is deposited upon the bottom of the sea, or in the estuary of rivers ; take it to be an inch, or two, or three inches a year, or whatever you may roughly estimate it at ; then take the total thickness of the whole series of stratified rocks, which geologists estimate at twelve or thirteen miles, or about seventy thousand feet, make a sum in short division, divide the total thickness by that of the quantity deposited in one year, and the result will, of course, give you the number of years which the crust has taken to form.

Truly, that looks a very simple process ! It would be so except for certain difficulties, the very first of which is that of finding how rapidly sediments are deposited ; but the main difficulty —a difficulty which renders any certain calculations of such a matter out of the question—is this, the sea-bottom on which the deposit takes place is continually shifting.

Instead of the surface of the earth being that stable, fixed thing that it is popularly believed to be, being, in common parlance, the very emblem of fixity itself, it is incessantly moving, and is, in fact, as unstable as the surface of the sea, except that its undulations are infinitely slower and enormously higher and deeper.

Now, what is the effect of this oscillation ? Take the case to which I have previously referred. The finer or coarser sediments that are carried down by the current of the river, will only be carried out a certain distance, and eventually, as we have already seen, on reaching the stiller part of the ocean, will be deposited at the bottom.

Let C y (Fig. 4) be the sea-bottom, y D the shore, x y the sea-level, then the coarser deposit will subside over the region B, the finer over A, while beyond A there will be no deposit at all;

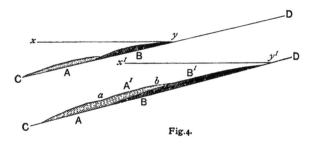

Fig. 4.

and, consequently, no record will be kept, simply because no deposit is going on. Now, suppose that the whole land, C, D, which we have regarded as stationary, goes down, as it does so, both A and B go further out from the shore, which will be at y^1; x^1, y^1, being the new sea-level. The consequence will be that the layer of mud (A), being now, for the most part, further than the force of the current is strong enough to convey even the finest *débris*, will, of course, receive no more deposits, and having attained a certain thickness will now grow no thicker.

We should be misled in taking the thickness of that layer, whenever it may be exposed to our view, as a record of time in the manner in which we are now regarding this subject, as it would give us only an imperfect and partial record: it would seem to represent too short a period of time.

Suppose, on the other hand, that the land (C D) had gone on rising slowly and gradually—say an inch or two inches in the course of a century,—what would be the practical effect of that movement? Why, that the sediment A and B which has been already deposited, would eventually be brought nearer to the shore-level and again subjected to the wear and tear of the sea; and directly the sea begins to act upon it, it would of course soon cut up and carry it way, to a greater or less extent, to be re-deposited further out.

Well, as there is, in all probability, not one single spot on the whole surface of the earth, which has not been up and down in this way a great many times, it follows that the thickness of the deposits formed at any particular spot cannot be taken (even supposing we had at first obtained correct data as to the rate at which they took place), as affording reliable information as to the period of time occupied in its deposit. So that you see it is absolutely necessary from these facts, seeing that

z 2

our record entirely consists of accumulations of mud, superimposed one on the other; seeing in the next place that any particular spots on which accumulations have occurred, have been constantly moving up and down, and sometimes out of the reach of a deposit, and at other times its own deposit broken up and carried away, it follows that our record must be in the highest degree imperfect, and we have hardly a trace left of thick deposits, or any definite knowledge of the area that they occupied, in a great many cases. And mark this! That supposing even that the whole surface of the earth had been accessible to the geologist,—that man had had access to every part of the earth, and had made sections of the whole, and put them all together,—even then his record must of necessity be imperfect.

But to how much has man really access? If you will look at this map you will see that it represents the proportion of the sea to the earth: this coloured part indicates all the dry land, and this other portion is the water. You will notice at once that the water covers three-fifths of the whole surface of the globe, and has covered it in the same manner ever since man has kept any record of his own observations, to say nothing of the minute period during which he has cultivated geological inquiry. So that three-fifths of the surface of the earth is shut out from us because it is under the sea. Let us look at the other

two-fifths, and see what are the countries in
which anything that may be termed searching
geological inquiry has been carried out : a good
deal of France, Germany, and Great Britain and
Ireland, bits of Spain, of Italy, and of Russia, have
been examined, but of the whole great mass of
Africa, except parts of the southern extremity,
we know next to nothing; little bits of India, but
of the greater part of the Asiatic continent
nothing; bits of the Northern American States
and of Canada, but of the greater part of the
continent of North America, and in still larger
proportion, of South America, nothing !

Under these circumstances, it follows that even
with reference to that kind of imperfect information which we can possess, it is only of about the
ten-thousandth part of the accessible parts of the
earth that has been examined properly.   Therefore, it is with justice that the most thoughtful of
those who are concerned in these inquiries insist
continually upon the imperfection of the geological
record ; for, I repeat, it is absolutely necessary,
from the nature of things, that that record should
be of the most fragmentary and imperfect
character.   Unfortunately this circumstance has
been constantly forgotten.   Men of science, like
young colts in a fresh pasture, are apt to be
exhilarated on being turned into a new field of
inquiry, to go off at a hand-gallop, in total
disregard of hedges and ditches, to lose sight of

the real limitation of their inquiries, and to forget the extreme imperfection of what is really known. Geologists have imagined that they could tell us what was going on at all parts of the earth's surface during a given epoch; they have talked of this deposit being contemporaneous with that deposit, until, from our little local histories of the changes at limited spots of the earth's surface, they have constructed a universal history of the globe as full of wonders and portents as any other story of antiquity.

But what does this attempt to construct a universal history of the globe imply? It implies that we shall not only have a precise knowledge of the events which have occurred at any particular point, but that we shall be able to say what events, at any one spot, took place at the same time with those at other spots.

Let us see how far that is in the nature of things practicable. Suppose that here I make a section of the Lake of Killarney, and here the section of another lake—that of Loch Lomond in Scotland for instance. The rivers that flow into them are constantly carrying down deposits of mud, and beds, or strata, are being as constantly formed, one above the other, at the bottom of those lakes. Now, there is not a shadow of doubt that in these two lakes the lower beds are all older than the upper—there is no doubt about that; but what does *this* tell us about the age of

any given bed in Loch Lomond, as compared with
that of any given bed in the Lake of Killarney?
It is, indeed, obvious that if any two sets of
deposits are separated and discontinuous, there is
absolutely no means whatever given you by the
nature of the deposit of saying whether one is
much younger or older than the other; but you
may say, as many have said and think, that the
case is very much altered if the beds which we
are comparing are continuous.   Suppose two beds

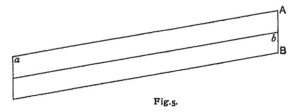

Fig. 5.

of mud hardened into rock,—A and B—are seen
in section.   (Fig. 5.)

Well, you say, it is admitted that the lower-
most bed is always the older.   Very well; B,
therefore, is older than A.   No doubt, *as a whole*,
it is so; or if any parts of the two beds which are
in the same vertical line are compared, it is so.
But suppose you take what seems a very natural
step further, and say that the part *a* of the bed A
is younger than the part *b* of the bed B.   Is this
sound reasoning?   If you find any record of
changes taking place at *b*, did they occur before

any events which took place while a was being deposited? It looks all very plain sailing, indeed, to say that they did; and yet there is no proof of anything of the kind. As the former Director of this Institution, Sir H. De la Beche, long ago showed, this reasoning may involve an entire fallacy. It is extremely possible that a may have been deposited ages before b. It is very easy to understand how that can be. To return to Fig. 4; when A and B were deposited, they were *substantially* contemporaneous; A being simply the finer deposit, and B the coarser of the same detritus or waste of land. Now suppose that that sea-bottom goes down (as shown in Fig. 4), so that the first deposit is carried no farther than a, forming the bed A^1, and the coarse no farther than b, forming the bed B^1, the result will be the formation of two continuous beds, one of fine sediment (A A^1) over-lapping another of coarse sediment (B B^1). Now suppose the whole sea-bottom is raised up, and a section exposed about the point A^1; no doubt, *at this spot*, the upper bed is younger than the lower. But we should obviously greatly err if we concluded that the mass of the upper bed at A was younger than the lower bed at B; for we have just seen that they are contemporaneous deposits. Still more should we be in error if we supposed the upper bed at A to be younger than the continuation of the lower bed at B^1; for A was deposited long before B^1.

In fine, if, instead of comparing immediately adjacent parts of two beds, one of which lies upon another, we compare distant parts, it is quite possible that the upper may be any number of years older than the under, and the under any number of years younger than the upper.

Now you must not suppose that I put this before you for the purpose of raising a paradoxical difficulty; the fact is, that the great mass of deposits have taken place in sea-bottoms which are gradually sinking, and have been formed under the very conditions I am here supposing.

Do not run away with the notion that this subverts the principle I laid down at first. The error lies in extending a principle which is perfectly applicable to deposits in the same vertical line to deposits which are not in that relation to one another.

It is in consequence of circumstances of this kind, and of others that I might mention to you, that our conclusions on and interpretations of the record are really and strictly only valid so long as we confine ourselves to one vertical section. I do not mean to tell you that there are no qualifying circumstances, so that, even in very considerable areas, we may safely speak of conformably superimposed beds being older or younger than others at many different points. But we can never be quite sure in coming to that conclusion, and especially we cannot be sure if there is any break in their continuity, or any very great distance between the points to be compared.

Well now, so much for the record itself,—so much for its imperfections,—so much for the conditions to be observed in interpreting it, and its chronological indications, the moment we pass beyond the limits of a vertical linear section.

Now let us pass from the record to that which it contains,—from the book itself to the writing and the figures on its pages. This writing and these figures consist of remains of animals and plants which, in the great majority of cases, have lived and died in the very spot in which we now find them, or at least in the immediate vicinity. You must all of you be aware—and I referred to the fact in my last lecture—that there are vast numbers of creatures living at the bottom of the sea. These creatures, like all others, sooner or later die, and their shells and hard parts lie at the bottom; and then the fine mud which is being constantly brought down by rivers and the action of the wear and tear of the sea, covers them over and protects them from any further change or alteration; and, of course, as in process of time the mud becomes hardened and solidified, the shells of these animals are preserved and firmly imbedded in the limestone or sandstone which is being thus formed. You may see in the galleries of the Museum up stairs specimens of limestones in which such fossil remains of existing animals are imbedded. There are some specimens in which turtles' eggs have been imbedded in calcareous sand, and before the sun had hatched the young turtles, they became covered over with calcareous mud, and thus have been preserved and fossilised.

Not only does this process of imbedding and fossilisation occur with marine and other aquatic animals and plants, but it affects those land animals and plants which are drifted away to sea, or become buried in bogs or morasses; and the animals which have been trodden down by their fellows and crushed in the mud at the river's bank, as the herd have come to drink. In any of these cases, the organisms may be crushed or be mutilated, before or after putrefaction, in such a manner that perhaps only a part will be left in the form in which it reaches us. It is, indeed, a most remarkable fact, that it is quite an exceptional case to find a skeleton of any one of all the thousands of wild land animals that we know are constantly being killed, or dying in the course of nature : they are preyed on and devoured by other animals, or die in places where their bodies are not afterwards protected by mud. There are other animals existing on the sea, the shells of which form exceedingly large deposits. You are probably aware that before the attempt was made to lay the Atlantic telegraphic cable, the Government employed vessels in making a series of very

careful observations and soundings of the bottom
of the Atlantic; and although, as we must all
regret, that up to the present time that project has
not succeeded, we have the satisfaction of knowing
that it yielded some most remarkable results to
science. The Atlantic Ocean had to be sounded
right across, to depths of several miles in some
places, and the nature of its bottom was carefully
ascertained. Well, now, a space of about 1,000
miles wide from east to west, and I do not exactly
know how many from north to south, but at any
rate 600 or 700 miles, was carefully examined, and
it was found that over the whole of that immense
area an excessively fine chalky mud is being
deposited; and this deposit is entirely made up of
animals whose hard parts are deposited in this
part of the ocean, and are doubtless gradually
acquiring solidity and becoming metamorphosed
into a chalky limestone. Thus, you see, it is quite
possible in this way to preserve unmistakable
records of animal and vegetable life. Whenever
the sea-bottom, by some of those undulations of
the earth's crust that I have referred to, becomes
up-heaved, and sections or borings are made, or
pits are dug, then we become able to examine
the contents and constituents of these ancient sea-
bottoms, and find out what manner of animals
lived at that period.

Now it is a very important consideration in its
bearing on the completeness of the record, to

inquire how far the remains contained in these
fossiliferous limestones are able to convey anything like an accurate or complete account of the
animals which were in existence at the time of its
formation. Upon that point we can form a very
clear judgment, and one in which there is no
possible room for any mistake. There are of
course a great number of animals—such as jelly-
fishes, and other animals—without any hard parts,
of which we cannot reasonably expect to find any
traces whatever: there is nothing of them to preserve. Within a very short time, you will have
noticed, after they are removed from the water,
they dry up to a mere nothing; certainly they
are not of a nature to leave any very visible traces
of their existence on such bodies as chalk or mud.
Then again, look at land animals; it is, as I have
said, a very uncommon thing to find a land animal
entire after death. Insects and other carnivorous
animals very speedily pull them to pieces, putrefaction takes place, and so, out of the hundreds of
thousands that are known to die every year, it is
the rarest thing in the world to see one imbedded
in such a way that its remains would be preserved
for a lengthened period. Not only is this the
case, but even when animal remains have been
safely imbedded, certain natural agents may wholly
destroy and remove them.

Almost all the hard parts of animals—the
bones and so on—are composed chiefly of phosphate

of lime and carbonate of lime. Some years ago, I had to make an inquiry into the nature of some very curious fossils sent to me from the North of Scotland. Fossils are usually hard bony structures that have become imbedded in the way I have described, and have gradually acquired the nature and solidity of the body with which they are associated; but in this case I had a series of *holes* in some pieces of rock, and nothing else. Those holes, however, had a certain definite shape about them, and when I got a skilful workman to make castings of the interior of these holes, I found that they were the impressions of the joints of a backbone and of the armour of a great reptile, twelve or more feet long. This great beast had died and got buried in the sand; the sand had gradually hardened over the bones, but remained porous. Water had trickled through it, and that water being probably charged with a superfluity of carbonic acid, had dissolved all the phosphate and carbonate of lime, and the bones themselves had thus decayed and entirely disappeared; but as the sandstone happened to have consolidated by that time, the precise shape of the bones was retained. If that sandstone had remained soft a little longer, we should have known nothing whatsoever of the existence of the reptile whose bones it had encased.

How certain it is that a vast number of animals which have existed at one period on this earth have entirely perished, and left no trace whatever of their forms, may be proved to you by other considerations. There are large tracts of sandstone in various parts of the world, in which nobody has yet found anything but footsteps. Not a bone of any description, but an enormous number of traces of footsteps. There is no question about them. There is a whole valley in Connecticut covered with these footsteps, and not a single fragment of the animals which made them have yet been found. Let me mention another case while upon that matter, which is even more surprising than those to which I have yet referred. There is a limestone formation near Oxford, at a place called Stonesfield, which has yielded the remains of certain very interesting mammalian animals, and up to this time, if I recollect rightly, there have been found seven specimens of its lower jaws, and not a bit of anything else, neither limb-bones nor skull, nor any part whatever; not a fragment of the whole system! Of course, it would be preposterous to imagine that the beasts had nothing else but a lower jaw! The probability is, as Dr. Buckland showed, as the result of his observations on dead dogs in the river Thames, that the lower jaw, not being secured by very firm ligaments to the bones of the head, and being a weighty affair, would easily be knocked off, or might drop away from the body as it floated in water in a state of de-

which will be requisite to consider carefully ; and
the first point for us is to examine how much the
extinct *Flora* and *Fauna* as a *whole*—disregarding
altogether the *succession* of their constituents, of
which I shall speak afterwards—differ from the
Flora and *Fauna* of the present day ;—how far they
differ in what we *do* know about them, leaving
altogether out of consideration speculations based
upon what we *do not* know.

I strongly imagine that if it were not for the
peculiar appearance that fossilised animals have,
any of you might readily walk through a
museum which contains fossil remains mixed up
with those of the present forms of life, and I doubt
very much whether your uninstructed eyes would
lead you to see any vast or wonderful difference
between the two.   If you looked closely, you would
notice, in the first place, a great many things very
like animals with which you are acquainted now :
you would see differences of shape and proportion,
but on the whole a close similarity.

I explained what I meant by ORDERS the other
day, when I described the animal kingdom as
being divided into sub-kingdoms, classes and
orders. If you divide the animal kingdom into
orders you will find that there are above one
hundred and twenty. The number may vary on
one side or the other, but this is a fair estimate.
That is the sum total of the orders of all the
animals which we know now, and which have

been known in past times, and left remains behind.

Now, how many of those are absolutely extinct? That is to say, how many of these orders of animals have lived at a former period of the world's history but have at present no representatives? That is the sense in which I meant to use the word "extinct." I mean that those animals did live on this earth at one time, but have left no one of their kind with us at the present moment. So that estimating the number of extinct animals is a sort of way of comparing the past creation as a whole with the present as a whole. Among the mammalia and birds there are none extinct; but when we come to the reptiles there is a most wonderful thing: out of the eight orders, or thereabouts, which you can make among reptiles, one-half are extinct. These diagrams of the plesiosaurus, the ichthyosaurus, the pterodactyle, give you a notion of some of these extinct reptiles. And here is a cast of the pterodactyle and bones of the ichthyosaurus and the plesiosaurus, just as fresh-looking as if it had been recently dug up in a churchyard. Thus, in the reptile class, there are no less than half of the orders which are absolutely extinct. If we turn to the *Amphibia*, there was one extinct order, the Labyrinthodonts, typified by the large salamander-like beast shown in this diagram.

No order of fishes is known to be extinct.

Every fish that we find in the strata—to which I have been referring—can be identified and placed in one of the orders which exist at the present day. There is not known to be a single ordinal form of insect extinct. There are only two orders extinct among the *Crustacea*. There is not known to be an extinct order of these creatures, the parasitic and other worms ; but there are two, not to say three, absolutely extinct orders of this class, the *Echinodermata ;* out of all the orders of the *Cœlenterata* and *Protozoa* only one, the Rugose Corals.

So that, you see, out of somewhere about 120 orders of animals, taking them altogether, you will not, at the outside estimate, find above ten or a dozen extinct. Summing up all the order of animals which have left remains behind them, you will not find above ten or a dozen which cannot be arranged with those of the present day ; that is to say, that the difference does not amount to much more than ten per cent.: and the proportion of extinct orders of plants is still smaller. I think that that is a very astounding a most astonishing fact : seeing the enormous epochs of time which have elapsed during the constitution of the surface of the earth as it at present exists, it is, indeed, a most astounding thing that the proportion of extinct ordinal types should be so exceedingly small.

But now, there is another point of view in which

we must look at this past creation. Suppose that we were to sink a vertical pit through the floor beneath us, and that I could succeed in making a section right through in the direction of New Zealand, I should find in each of the different beds through which I passed the remains of animals which I should find in that stratum and not in the others. First, I should come upon beds of gravel or drift containing the bones of large animals, such as the elephant, rhinoceros, and cave tiger. Rather curious things to fall across in Piccadilly! If I should dig lower still, I should come upon a bed of what we call the London clay, and in this, as you will see in our galleries up stairs, are found remains of strange cattle, remains of turtles, palms, and large tropical fruits; with shell-fish such as you see the like of now only in tropical regions. If I went below that, I should come upon the chalk, and there I should find something altogether different, the remains of ichthyosauria and pterodactyles, and ammonites, and so forth.

I do not know what Mr. Godwin Austin would say comes next, but probably rocks containing more ammonites, and more ichthyosauria and plesiosauria, with a vast number of other things; and under that I should meet with yet older rocks containing numbers of strange shells and fishes; and in thus passing from the surface to the lowest depths of the earth's crust, the forms of animal life and vegetable life which I should meet with in the successive beds would, looking at them broadly, be the more different the further that I went down. Or, in other words, inasmuch as we started with the clear principle, that in a series of naturally-disposed mud beds the lowest are the oldest, we should come to this result, that the further we go back in time the more difference exists between the animal and vegetable life of an epoch and that which now exists. That was the conclusion to which I wished to bring you at the end of this lecture.

III

THE METHOD BY WHICH THE CAUSES OF THE
PRESENT AND PAST CONDITIONS OF ORGANIC
NATURE ARE TO BE DISCOVERED.—THE
ORIGINATION OF LIVING BEINGS.

IN the two preceding lectures I have endeavoured
to indicate to you the extent of the subject-matter
of the inquiry upon which we are engaged; and
having thus acquired some conception of the past
and present phenomena of organic nature, I must
now turn to that which constitutes the great problem which we have set before ourselves;—I mean,
the question of what knowledge we have of the
causes of these phenomena of organic nature, and
how such knowledge is obtainable.

Here, on the threshold of the inquiry, an
objection meets us. There are in the world a
number of extremely worthy, well-meaning
persons, whose judgments and opinions are
entitled to the utmost respect on account of
their sincerity, who are of opinion that vital

phenomena, and especially all questions relating to the origin of vital phenomena, are questions quite apart from the ordinary run of inquiry, and are, by their very nature, placed out of our reach. They say that all these phenomena originated miraculously, or in some way totally different from the ordinary course of nature, and that therefore they conceive it to be futile, not to say presumptuous, to attempt to inquire into them.

To such sincere and earnest persons, I would only say, that a question of this kind is not to be shelved upon theoretical or speculative grounds. You may remember the story of the Sophist who demonstrated to Diogenes in the most complete and satisfactory manner that he could not walk ; that, in fact, all motion was an impossibility ; and that Diogenes refuted him by simply getting up and walking round his tub.   So, in the same way, the man of science replies to objections of this kind, by simply getting up and walking onward, and showing what science has done and is doing —by pointing to that immense mass of facts which have been ascertained as systematised under the forms of the great doctrines of morphology, of development, of distribution, and the like.   He sees an enormous mass of facts and laws relating to organic beings, which stand on the same good sound foundation as every other natural law.   With this mass of facts and laws before us, therefore, seeing that, as far as organic matters rate, setting out to discover how much we at present know upon these abstruse matters, the question arises as to what is to be our course of proceeding, and what method we must lay down for our guidance. I reply to that question, that our method must be exactly the same as that which is pursued in any other scientific inquiry, the method of scientific investigation being the same for all orders of facts and phenomena whatsoever.

I must dwell a little on this point, for I wish you to leave this room with a very clear conviction that scientific investigation is not, as many people seem to suppose, some kind of modern black art. I say that you might easily gather this impression from the manner in which many persons speak of scientific inquiry, or talk about inductive and deductive philosophy, or the principles of the " Baconian philosophy." I do protest that, of the vast number of cants in this world, there are none, to my mind, so contemptible as the pseudo-scientific cant which is talked about the " Baconian philosophy."

To hear people talk about the great Chancellor —and a very great man he certainly was,—you would think that it was he who had invented science, and that there was no such thing as sound reasoning before the time of Queen Elizabeth! Of course you say, that cannot possibly be true; you perceive, on a moment's reflection, that such an idea is absurdly wrong, and yet, so firmly rooted is this sort of impression, —I cannot call it an idea, or conception,—the thing is too absurd to be entertained,—but so completely does it exist at the bottom of most men's minds, that this has been a matter of observation with me for many years past. There are many men who, though knowing absolutely nothing of the subject with which they may be dealing, wish, nevertheless, to damage the author of some view with which they think fit to disagree. What they do, then, is not to go and learn something about the subject, which one would naturally think the best way of fairly dealing with it ; but they abuse the originator of the view they question, in a general manner, and wind up by saying that, "After all, you know, the principles and method of this author are totally opposed to the canons of the Baconian philosophy." Then everybody applauds, as a matter of course, and agrees that it must be so. But if you were to stop them all in the middle of their applause, you would probably find that neither the speaker nor his applauders could tell you how or in what way it was so ; neither the one nor the other having the slightest idea of what they mean when they speak of the " Baconian philosophy."

You will understand, I hope, that I have not the slightest desire to join in the outcry against either the morals, the intellect, or the great genius of Lord Chancellor Bacon. He was undoubtedly

a very great man, let people say what they will of
him; but notwithstanding all that he did for
philosophy, it would be entirely wrong to suppose
that the methods of modern scientific inquiry
originated with him, or with his age; they originated with the first man, whoever he was; and
indeed existed long before him, for many of the
essential processes of reasoning are exerted by the
higher order of brutes as completely and effectively
as by ourselves. We see in many of the brute
creation the exercise of one, at least, of the same
powers of reasoning as that which we ourselves
employ.

The method of scientific investigation is nothing
but the expression of the necessary mode of working of the human mind. It is simply the mode
at which all phenomena are reasoned about, rendered precise and exact. There is no more difference, but there is just the same kind of difference,
between the mental operations of a man of science
and those of an ordinary person, as there is between
the operations and methods of a baker or of a
butcher weighing out his goods in common scales,
and the operations of a chemist in performing a
difficult and complex analysis by means of his
balance and finely-graduated weights. It is not
that the action of the scales in the one case, and
the balance in the other, differ in the principles of
their construction or manner of working; but the
beam of one is set on an infinitely finer axis than

the other, and of course turns by the addition of a much smaller weight.

You will understand this better, perhaps, if I give you some familiar example. You have all heard it repeated, I dare say, that men of science work by means of induction and deduction, and that by the help of these operations, they, in a sort of sense, wring from Nature certain other things, which are called natural laws, and causes, and that out of these, by some cunning skill of their own, they build up hypotheses and theories. And it is imagined by many, that the operations of the common mind can be by no means compared with these processes, and that they have to be acquired by a sort of special apprenticeship to the craft. To hear all these large words, you would think that the mind of a man of science must be constituted differently from that of his fellow men ; but if you will not be frightened by terms, you will discover that you are quite wrong, and that all these terrible apparatus are being used by yourselves every day and every hour of your lives.

There is a well-known incident in one of Molière's plays, where the author makes the hero express unbounded delight on being told that he had been talking prose during the whole of his life. In the same way, I trust, that you will take comfort, and be delighted with yourselves, on the discovery that you have been acting on the prin-

ciples of inductive and deductive philosophy during the same period. Probably there is not one
here who has not in the course of the day had
occasion to set in motion a complex train of reasoning, of the very same kind, though differing of
course in degree, as that which a scientific man
goes through in tracing the causes of natural
phenomena.

A very trivial circumstance will serve to exemplify this. Suppose you go into a fruiterer's
shop, wanting an apple,—you take up one, and,
on biting it, you find it is sour; you look at it,
and see that it is hard and green. You take
up another one, and that too is hard, green,
and sour. The shopman offers you a third;
but, before biting it, you examine it, and find
that it is hard and green, and you immediately
say that you will not have it, as it must
be sour, like those that you have already
tried.

Nothing can be more simple than that, you
think; but if you will take the trouble to analyse
and trace out into its logical elements what has
been done by the mind, you will be greatly surprised. In the first place, you have performed
the operation of induction. You found that, in
two experiences, hardness and greenness in apples
went together with sourness. It was so in the
first case, and it was confirmed by the second.
True, it is a very small basis, but still it is enough

to make an induction from; you generalise the facts, and you expect to find sourness in apples where you get hardness and greenness. You found upon that a general law, that all hard and green apples are sour; and that, so far as it goes, is a perfect induction. Well, having got your natural law in this way, when you are offered another apple which you find is hard and green, you say, "All hard and green apples are sour; this apple is hard and green, therefore this apple is sour." That train of reasoning is what logicians call a syllogism, and has all its various parts and terms, —its major premiss, its minor premiss, and its conclusion. And, by the help of further reasoning, which, if drawn out, would have to be exhibited in two or three other syllogisms, you arrive at your final determination, "I will not have that apple." So that, you see, you have, in the first place, established a law by induction, and upon that you have founded a deduction, and reasoned out the special conclusion of the particular case. Well now, suppose, having got your law, that at some time afterwards, you are discussing the qualities of apples with a friend : you will say to him, "It is a very curious thing,—but I find that all hard and green apples are sour!" Your friend says to you, "But how do you know that?" You at once reply, "Oh, because I have tried them over and over again, and have always found them to be so." Well, if we were talking science instead of common sense, we should call that an experimental verification. And, if still opposed, you go further, and say, "I have heard from the people in Somersetshire and Devonshire, where a large number of apples are grown, that they have observed the same thing. It is also found to be the case in Normandy, and in North America. In short, I find it to be the universal experience of mankind wherever attention has been directed to the subject." Whereupon, your friend, unless he is a very unreasonable man, agrees with you, and is convinced that you are quite right in the conclusion you have drawn. He believes, although perhaps he does not know he believes it, that the more extensive verifications are,—that the more frequently experiments have been made, and results of the same kind arrived at,—that the more varied the conditions under which the same results are attained, the more certain is the ultimate conclusion, and he disputes the question no further. He sees that the experiment has been tried under all sorts of conditions, as to time, place, and people, with the same result; and he says with you, therefore, that the law you have laid down must be a good one, and he must believe it.

In science we do the same thing;—the philosopher exercises precisely the same faculties, though in a much more delicate manner. In scientific inquiry it becomes a matter of duty to expose a supposed law to every possible kind of verification, and to take care, moreover, that this is done intentionally, and not left to a mere accident, as in the case of the apples. And in science, as in common life, our confidence in a law is in exact proportion to the absence of variation in the result of our experimental verifications. For instance, if you let go your grasp of an article you may have in your hand, it will immediately fall to the ground. That is a very common verification of one of the best established laws of nature—that of gravitation. The method by which men of science establish the existence of that law is exactly the same as that by which we have established the trivial proposition about the sourness of hard and green apples. But we believe it in such an extensive, thorough, and unhesitating manner because the universal experience of mankind verifies it, and we can verify it ourselves at any time; and that is the strongest possible foundation on which any natural law can rest.

So much, then, by way of proof that the method of establishing laws in science is exactly the same as that pursued in common life. Let us now turn to another matter (though really it is but another phase of the same question), and that is, the method by which, from the relations of certain phenomena, we prove that some stand in the position of causes towards the others.

I want to put the case clearly before you, and I will therefore show you what I mean by another

familiar example. I will suppose that one of you,
on coming down in the morning to the parlour of
your house, finds that a tea-pot and some spoons
which had been left in the room on the previous
evening are gone,—the window is open, and you
observe the mark of a dirty hand on the window-
frame, and perhaps, in addition to that, you notice
the impress of a hob-nailed shoe on the gravel
outside. All these phenomena have struck your
attention instantly, and before two seconds have
passed you say, " Oh, somebody has broken open
the window, entered the room, and run off with
the spoons and the tea-pot ! " That speech is out
of your mouth in a moment. And you will probably add, " I know there has ; I am quite sure of
it ! " You mean to say exactly what you know ;
but in reality you are giving expression to what
is, in all essential particulars, an hypothesis.
You do not *know* it at all ; it is nothing but an
hypothesis rapidly framed in your own mind. And
it is an hypothesis founded on a long train of inductions and deductions.

What are those inductions and deductions, and
how have you got at this hypothesis ? You have
observed, in the first place, that the window is
open ; but by a train of reasoning involving many
inductions and deductions, you have probably
arrived long before at the general law—and a
very good one it is—that windows do not open of
themselves ; and you therefore conclude that

something has opened the window. A second general law that you have arrived at in the same way is, that tea-pots and spoons do not go out of a window spontaneously, and you are satisfied that, as they are not now where you left them, they have been removed. In the third place, you look at the marks on the window-sill, and the shoe-marks outside, and you say that in all previous experience the former kind of mark has never been produced by anything else but the hand of a human being ; and the same experience shows that no other animal but man at present wears shoes with hob-nails in them such as would produce the marks in the gravel. I do not know, even if we could discover any of those " missing links " that are talked about, that they would help us to any other conclusion ! At any rate the law which states our present experience is strong enough for my present purpose. You next reach the con-clusion, that as these kinds of marks have not been left by any other animals than men, or are liable to be formed in any other way than by a man's hand and shoe, the marks in question have been formed by a man in that way. You have, further, a general law, founded on observation and experi-ence, and that, too, is, I am sorry to say, a very universal and unimpeachable one,—that some men are thieves ; and you assume at once from all these premisses—and that is what constitutes your hypothesis—that the man who made the marks outside and on the window-sill, opened the window, got into the room, and stole your tea-pot and spoons. You have now arrived at a *vera causa;* —you have assumed a cause which, it is plain, is competent to produce all the phenomena you have observed. You can explain all these phenomena only by the hypothesis of a thief. But that is a hypothetical conclusion, of the justice of which you have no absolute proof at all; it is only rendered highly probable by a series of inductive and deductive reasonings.

I suppose your first action, assuming that you are a man of ordinary common sense, and that you have established this hypothesis to your own satisfaction, will very likely be to go off for the police, and set them on the track of the burglar, with the view to the recovery of your property. But just as you are starting with this object, some person comes in, and on learning what you are about, says, "My good friend, you are going on a great deal too fast. How do you know that the man who really made the marks took the spoons? It might have been a monkey that took them, and the man may have merely looked in afterwards." You would probably reply, "Well, that is all very well, but you see it is contrary to all experience of the way tea-pots and spoons are abstracted; so that, at any rate, your hypothesis is less probable than mine." While you are talking the thing over in this way, another friend arrives, one of that good kind of people that I was talking of a little while ago. And he might say, " Oh, my dear sir, you are certainly going on a great deal too fast. You are most presumptuous. You admit that all these occurrences took place when you were fast asleep, at a time when you could not possibly have known anything about what was taking place. How do you know that the laws of Nature are not suspended during the night? It may be that there has been some kind of supernatural interference in this case." In point of fact, he declares that your hypothesis is one of which you cannot at all demonstrate the truth, and that you are by no means sure that the laws of Nature are the same when you are asleep as when you are awake.

Well, now, you cannot at the moment answer that kind of reasoning. You feel that your worthy friend has you somewhat at a disadvantage. You will feel perfectly convinced in your own mind, however, that you are quite right, and you say to him, " My good friend, I can only be guided by the natural probabilities of the case, and if you will be kind enough to stand aside and permit me to pass, I will go and fetch the police." Well, we will suppose that your journey is successful, and that by good luck you meet with a policeman; that eventually the burglar is found with your property on his person, and the marks correspond to his hand and to his boots. Probably any jury would consider those facts a very good experimental verification of your hypothesis, touching the cause of the abnormal phenomena observed in your parlour, and would act accordingly.

Now, in this suppositious case, I have taken phenomena of a very common kind, in order that you might see what are the different steps in an ordinary process of reasoning, if you will only take the trouble to analyse it carefully. All the operations I have described, you will see, are involved in the mind of any man of sense in leading him to a conclusion as to the course he should take in order to make good a robbery and punish the offender. I say that you are led, in that case, to your conclusion by exactly the same train of reasoning as that which a man of science pursues when he is endeavouring to discover the origin and laws of the most occult phenomena. The process is, and always must be, the same; and precisely the same mode of reasoning was employed by Newton and Laplace in their endeavours to discover and define the causes of the movements of the heavenly bodies, as you, with your own common sense, would employ to detect a burglar. The only difference is, that the nature of the inquiry being more abstruse, every step has to be most carefully watched, so that there may not be a single crack or flaw in your hypothesis. A flaw or crack in many of the hypotheses of daily life may be of little or no moment as affecting the general correctness of the conclusions at which we may arrive; but, in a scientific inquiry, a fallacy, great or small, is always of importance, and is sure to be in the long run constantly productive of mischievous, if not fatal results.

Do not allow yourselves to be misled by the common notion that an hypothesis is untrustworthy simply because it is an hypothesis. It is often urged, in respect to some scientific conclusion, that, after all, it is only an hypothesis. But what more have we to guide us in nine-tenths of the most important affairs of daily life than hypotheses, and often very ill-based ones ? So that in science, where the evidence of an hypothesis is subjected to the most rigid examination, we may rightly pursue the same course. You may have hypotheses and hypotheses. A man may say, if he likes, that the moon is made of green cheese: that is an hypothesis. But another man, who has devoted a great deal of time and attention to the subject, and availed himself of the most powerful telescopes and the results of the observations of others, declares that in his opinion it is probably composed of materials very similar to those of which our own earth is made up : and that is also only an hypothesis. But I need not tell you that there is an enormous difference in the value of the

two hypotheses. That one which is based on
sound scientific knowledge is sure to have a corresponding value; and that which is a mere hasty
random guess is likely to have but little value.
Every great step in our progress in discovering
causes has been made in exactly the same way as
that which I have detailed to you. A person
observing the occurrence of certain facts and
phenomena asks, naturally enough, what process,
what kind of operation known to occur in Nature
applied to the particular case, will unravel and
explain the mystery? Hence you have the
scientific hypothesis; and its value will be pro-
portionate to the care and completeness with which
its basis had been tested and verified. It is in
these matters as in the commonest affairs of practical life: the guess of the fool will be folly, while
the guess of the wise man will contain wisdom.
In all cases, you see that the value of the result
depends on the patience and faithfulness with
which the investigator applies to his hypothesis
every possible kind of verification.

I dare say I may have to return to this point
by and by; but having dealt thus far with our
logical methods, I must now turn to something
which, perhaps, you may consider more interesting,
or, at any rate, more tangible. But in reality
there are but few things that can be more important for you to understand than the mental processes and the means by which we obtain scientific

conclusions and theories.[1] Having granted that
the inquiry is a proper one, and having determined
on the nature of the methods we are to pursue
and which only can lead to success, I must now
turn to the consideration of our knowledge of the
nature of the processes which have resulted in the
present condition of organic nature.

Here, let me say at once, lest some of you misunderstand me, that I have extremely little to
report. The question of how the present condition
of organic nature came about, resolves itself into
two questions. The first is: How has organic or
living matter commenced its existence? And the
second is: How has it been perpetuated? On the
second question I shall have more to say hereafter.
But on the first one, what I now have to say will
be for the most part of a negative character.

If you consider what kind of evidence we can
have upon this matter, it will resolve itself into
two kinds. We may have historical evidence and we
may have experimental evidence. It is, for example,
conceivable, that inasmuch as the hardened mud
which forms a considerable portion of the thickness of the earth's crust contains faithful records
of the past forms of life, and inasmuch as these
differ more and more as we go further down,—it
is possible and conceivable that we might come to

---

[1] Those who wish to study fully the doctrines of which I
have endeavoured to give some rough-and-ready illustrations,
must read Mr. John Stuart Mill's *System of Logic*.

some particular bed or stratum which should contain the remains of those creatures with which
organic life began upon the earth.   And if we did
so, and if such forms of organic life were pre-
servable, we should have what I would call historical evidence of the mode in which organic life
began upon this planet. Many persons will tell
you, and indeed you will find it stated in many
works on geology, that this has been done, and
that we really possess such a record; there are
some who imagine that the earliest forms of life
of which we have as yet discovered any record, are
in truth the forms in which animal life began upon
the globe.   The grounds on which they base that
supposition are these :—That if you go through
the enormous thickness of the earth's crust and
get down to the older rocks, the higher vertebrate
animals—the quadrupeds, birds, and fishes—cease
to be found; beneath them you find only the invertebrate animals ; and in the deepest and lowest
rocks those remains become scantier and scantier,
not in any very gradual progression, however,
until, at length, in what are supposed to be the
oldest rocks, the animal remains which are found
are almost always confined to four forms—*Oldhamia*,
whose precise nature is not known, whether plant
or animal; *Lingula,* a kind of mollusc ; *Trilobites,*
a crustacean animal, having the same essential
plan of construction, though differing in many
details from a lobster or crab ; and *Hymenocaris,*

evidence we have there. To enable us to say that
we know anything about the experimental origination of organisation and life, the investigator
ought to be able to take inorganic matters, such
as carbonic acid, ammonia, water, and salines, in
any sort of inorganic combination, and be able to
build them up into protein matter, and then that
protein matter ought to begin to live in an
organic form.   That, nobody has done as yet, and
I suspect it will be a long while before anybody
does do it. But the thing is by no means so
impossible as it looks ; for the researches of modern
chemistry have shown us—I won't say the road
towards it, but, if I may so say, they have shown
the finger-post pointing to the road that may lead
to it.

It is not many years ago—and you must recollect that Organic Chemistry is a young science,
not above a couple of generations old, you must
not expect too much of it,—it is not many years
ago since it was said to be perfectly impossible to
fabricate any organic compound ; that is to say,
any non-mineral compound which is to be found
in an organised being.   It remained so for a very
long period ; but it is now a considerable number
of years since a distinguished foreign chemist contrived to fabricate urea, a substance of a very
complex character, which forms one of the waste
products of animal structures.   And of late years
a number of other compounds, such as butyric

acid, and others, have been added to the list. I
need not tell you that chemistry is an enormous
distance from the goal I indicate; all I wish to
point out to you is, that it is by no means safe
to say that that goal may not be reached one
day. It may be that it is impossible for us
to produce the conditions requisite to the origination of life; but we must speak modestly about
the matter, and recollect that Science has put her
foot upon the bottom round of the ladder. Truly
he would be a bold man who would venture to
predict where she will be fifty years hence.

There is another inquiry which bears indirectly
upon this question, and upon which I must say a
few words. You are all of you aware of the
phenomena of what is called spontaneous generation. Our forefathers, down to the seventeenth
century, or thereabouts, all imagined, in perfectly
good faith, that certain vegetable and animal
forms gave birth, in the process of their decomposition, to insect life. Thus, if you put a piece
of meat in the sun, and allowed it to putrefy, they
conceived that the grubs which soon began to
appear were the result of the action of a power of
spontaneous generation which the meat contained.
And they could give you receipts for making
various animal and vegetable preparations which
would produce particular kinds of animals. A
very distinguished Italian naturalist, named Redi,
took up the question, at a time when everybody

believed in it; among others our own great Harvey, the discoverer of the circulation of the blood. You will constantly find his name quoted, however, as an opponent of the doctrine of spontaneous generation; but the fact is, and you will see it if you will take the trouble to look into his works, Harvey believed it as profoundly as any man of his time; but he happened to enunciate a very curious proposition—that every living thing came from an *egg;* he did not mean to use the word in the sense in which we now employ it, he only meant to say that every living thing originated in a little rounded particle of organised substance; and it is from this circumstance, probably, that the notion of Harvey having opposed the doctrine originated. Then came Redi, and he proceeded to upset the doctrine in a very simple manner. He merely covered the piece of meat with some very fine gauze, and then he exposed it to the same conditions. The result of this was that no grubs or insects were produced; he proved that the grubs originated from the insects who came and deposited their eggs in the meat, and that they were hatched by the heat of the sun. By this kind of inquiry he thoroughly upset the doctrine of spontaneous generation, for his time at least.

Then came the discovery and application of the microscope to scientific inquiries, which showed to naturalists that besides the organisms which they already knew as living beings and plants, there were an immense number of minute things which could be obtained apparently almost at will from decaying vegetable and animal forms. Thus, if you took some ordinary black pepper or some hay, and steeped it in water, you would find in the course of a few days that the water had become impregnated with an immense number of animalcules swimming about in all directions. From facts of this kind naturalists were led to revive the theory of spontaneous generation. They were headed here by an English naturalist,—Needham,—and afterwards in France by the learned Buffon. They said that these things were absolutely begotten in the water of the decaying substances out of which the infusion was made. It did not matter whether you took animal or vegetable matter, you had only to steep it in water and expose it, and you would soon have plenty of animalcules. They made an hypothesis about this which was a very fair one. They said, this matter of the animal world, or of the higher plants, appears to be dead, but in reality it has a sort of dim life about it, which, if it is placed under fair conditions, will cause it to break up into the forms of these little animalcules, and they will go through their lives in the same way as the animal or plant of which they once formed a part.

The question now became very hotly debated. Spallanzani, an Italian naturalist, took up opposite

views to those of Needham and Buffon, and by
means of certain experiments he showed that it
was quite possible to stop the process by boiling
the water, and closing the vessel in which it was
contained. "Oh!" said his opponents; "but what
do you know you may be doing when you heat the
air over the water in this way? You may be de-
stroying some property of the air requisite for the
spontaneous generation of the animalcules."

However, Spallanzani's views were supposed to
be upon the right side, and those of the others fell
into discredit; although the fact was that Spallanzani had not made good his views. Well, then,
the subject continued to be revived from time to
time, and experiments were made by several persons; but these experiments were not altogether
satisfactory. It was found that if you put an infusion in which animalcules would appear if it were
exposed to the air into a vessel and boiled it, and
then sealed up the mouth of the vessel, so that no
air, save such as had been heated to 212°, could
reach its contents, that then no animalcules would
be found; but if you took the same vessel and exposed the infusion to the air, then you would get
animalcules. Furthermore, it was found that if
you connected the mouth of the vessel with a red-
hot tube in such a way that the air would have to
pass through the tube before reaching the infusion,
that then you would get no animalcules. Yet
another thing was noticed: if you took two flasks

containing the same kind of infusion, and left one
entirely exposed to the air, and in the mouth of
the other placed a ball of cotton wool, so that the
air would have to filter itself through it before
reaching the infusion, that then, although you
might have plenty of animalcules in the first flask,
you would certainly obtain none from the second.

These experiments, you see, all tended towards
one conclusion—that the infusoria were developed
from little minute spores or eggs which were constantly floating in the atmosphere, and which lose
their power of germination if subjected to heat.
But one observer now made another experiment,
which seemed to go entirely the other way, and
puzzled him altogether. He took some of this
boiled infusion that I have been speaking of, and
by the use of a mercurial bath—a kind of trough
used in laboratories—he deftly inverted a vessel
containing the infusion into the mercury, so that
the latter reached a little beyond the level of the
mouth of the *inverted* vessel. You see that he
thus had a quantity of the infusion shut off from
any possible communication with the outer air by
being inverted upon a bed of mercury.

He then prepared some pure oxygen and nitrogen gases, and passed them by means of a tube
going from the outside of the vessel, up through
the mercury into the infusion; so that he thus
had it exposed to a perfectly pure atmosphere of
the same constituents as the external air. Of

course, he expected he would get no infusorial animalcules at all in that infusion; but, to his great dismay and discomfiture, he found he almost always did get them.

Furthermore, it has been found that experiments made in the manner described above answer well with most infusions; but that if you fill the vessel with boiled milk, and then stop the neck with cotton-wool, you *will* have infusoria. So that you see there were two experiments that brought you to one kind of conclusion, and three to another; which was a most unsatisfactory state of things to arrive at in a scientific inquiry.

Some few years after this, the question began to be very hotly discussed in France. There was M. Pouchet, a professor at Rouen, a very learned man, but certainly not a very rigid experimentalist. He published a number of experiments of his own, some of which were very ingenious, to show that if you went to work in a proper way, there was a truth in the doctrine of spontaneous generation. Well, it was one of the most fortunate things in the world that M. Pouchet took up this question, because it induced a distinguished French chemist, M. Pasteur, to take up the question on the other side; and he has certainly worked it out in the most perfect manner. I am glad to say, too, that he has published his researches in time to enable me to give you an account of them. He verified all the experiments which I have just mentioned

to you—and then finding those extraordinary
anomalies, as in the case of the mercury bath and
the milk, he set himself to work to discover their
nature.   In the case of milk he found it to be a
question of temperature.   Milk in a fresh state is
slightly alkaline ; and it is a very curious circumstance, but this very slight degree of alkalinity
seems to have the effect of preserving the organisms which fall into it from the air from being
destroyed at a temperature of 212, which is the
boiling point.   But if you raise the temperature
10° when you boil it, the milk behaves like everything else ; and if the air with which it comes in
contact, after being boiled at this temperature, is
passed through a red-hot tube, you will not get a
trace of organisms.

He then turned his attention to the mercury
bath, and found on examination that the surface of
the mercury was almost always covered with a
very fine dust.   He found that even the mercury
itself was positively full of organic matters ; that
from being constantly exposed to the air, it had
collected an immense number of these infusorial
organisms from the air. Well, under these circumstances he felt that the case was quite clear, and
that the mercury was not what it had appeared to
M. Schwann to be,—a bar to the admission of these
organisms ; but that, in reality, it acted as a reservoir
from which the infusion was immediately supplied
with the large quantity that had so puzzled him.

But not content with explaining the experiments of others, M. Pasteur went to work to satisfy himself completely. He said to himself: "If my view is right, and if, in point of fact, all these appearances of spontaneous generation are altogether due to the falling of minute germs suspended in the atmosphere,—why, I ought not only to be able to show the germs, but I ought to be able to catch and sow them, and produce the resulting organisms." He, accordingly, constructed a very ingenious apparatus to enable him to accomplish the trapping of the "*germ dust*" in the air. He fixed in the window of his room a glass tube, in the centre of which he had placed a ball of gun-cotton, which, as you all know, is ordinary cotton-wool, which, from having been steeped in strong acid, is converted into a substance of great explosive power. It is also soluble in alcohol and ether. One end of the glass tube was, of course, open to the external air; and at the other end of it he placed an aspirator, a contrivance for causing a current of the external air to pass through the tube. He kept this apparatus going for four-and-twenty hours, and then removed the *dusted* gun-cotton, and dissolved it in alcohol and ether. He then allowed this to stand for a few hours, and the result was, that a very fine dust was gradually deposited at the bottom of it. That dust, on being transferred to the stage of a microscope, was found to contain an enormous number of starch grains.

taneous generation. He had succeeded in catching the germs and developing organisms in the way he had anticipated.

It now struck him that the truth of his conclusions might be demonstrated without all the apparatus he had employed. To do this, he took some decaying animal or vegetable substance, such as urine, which is an extremely decomposable substance, or the juice of yeast, or perhaps some other artificial preparation, and filled a vessel having a long tubular neck with it. He then boiled the liquid and bent that long neck into an S shape or zig-zag, leaving it open at the end. The infusion then gave no trace of any appearance of spontaneous generation, however long it might be left, as all the germs in the air were deposited in the beginning of the bent neck. He then cut the tube close to the vessel, and allowed the ordinary air to have free and direct access ; and the result of that was the appearance of organisms in it, as soon as the infusion had been allowed to stand long enough to allow of the growth of those it received from the air, which was about forty-eight hours. The result of M. Pasteur's experiments proved, therefore, in the most conclusive manner, that all the appearances of spontaneous generation arose from nothing more than the deposition of the germs of organisms which were constantly floating in the air.

To this conclusion, however, the objection was made, that if that were the cause, then the air would contain such an enormous number of these germs, that it would be a continual fog. But M. Pasteur replied that they are not there in anything like the number we might suppose, and that an exaggerated view has been held on that subject; he showed that the chances of animal or vegetable life appearing in infusions, depend entirely on the conditions under which they are exposed. If they are exposed to the ordinary atmosphere around us, why, of course, you may have organisms appearing early. But, on the other hand, if they are exposed to air at a great height, or in some very quiet cellar, you will often not find a single trace of life.

So that M. Pasteur arrived at last at the clear and definite result, that all these appearances are like the case of the worms in the piece of meat, which was refuted by Redi, simply germs carried by the air and deposited in the liquids in which they afterwards appear. For my own part, I conceive that, with the particulars of M. Pasteur's experiments before us, we cannot fail to arrive at his conclusions; and that the doctrine of spontaneous generation has received a final *coup de grâce*.

You, of course, understand that all this in no way interferes with the *possibility* of the fabrication of organic matters by the direct method to which I have referred, remote as that possibility may be.

IV

THE PERPETUATION OF LIVING BEINGS, HEREDITARY TRANSMISSION AND VARIATION.

THE inquiry which we undertook, at our last meeting, into the state of our knowledge of the causes of the phenomena of organic nature,—of the past and of the present,—resolved itself into two subsidiary inquiries : the first was, whether we know anything, either historically or experimentally, of the mode of origin of living beings; the second subsidiary inquiry was, whether, granting the origin, we know anything about the perpetuation and modifications of the forms of organic beings. The reply which I had to give to the first question was altogether negative, and the chief result of my last lecture was, that, neither historically nor experimentally, do we at present know anything whatsoever about the origin of living forms. We saw that, historically, we are not likely to know anything about it, although we may perhaps learn something experimentally ; but that at present we are an enormous distance from the goal I indicated.

I now, then, take up the next question, What
do we know of the reproduction, the perpetuation,
and the modifications of the forms of living beings,
supposing that we have put the question as to their
origination on one side, and have assumed that at
present the causes of their origination are beyond
us, and that we know nothing about them ? Upon
this question the state of our knowledge is extremely different ; it is exceedingly large : and, if
not complete, our experience is certainly most extensive.   It would be impossible to lay it all before
you, and the most I can do, or need do to-night, is
to take up the principal points and put them before you with such prominence as may subserve
the purposes of our present argument.

The method of the perpetuation of organic beings
is of two kinds,—the non-sexual and the sexual. In
the first the perpetuation takes place from and by
a particular act of an individual organism, which
sometimes may not be classed as belonging to any
sex at all.   In the second case, it is in consequence of the mutual action and interaction of
certain portions of the organisms of usually two
distinct individuals,—the male and the female. The
cases of non-sexual perpetuation are by no means
so common as the cases of sexual perpetuation ;
and they are by no means so common in the animal
as in the vegetable world.   You are all probably
familiar with the fact, as a matter of experience,
that you can propagate plants by means of what

are called "cuttings"; for example, that by taking a cutting from a geranium plant, and rearing it properly, by supplying it with light and warmth and nourishment from the earth, it grows up and takes the form of its parent, having all the properties and peculiarities of the original plant.

Sometimes this process, which the gardener performs artificially, takes place naturally; that is to say, a little bulb, or portion of the plant, detaches itself, drops off, and becomes capable of growing as a separate thing. That is the case with many bulbous plants, which throw off in this way secondary bulbs, which are lodged in the ground and become developed into plants. This is a non-sexual process, and from it results the repetition or reproduction of the form of the original being from which the bulb proceeds.

Among animals the same thing takes place. Among the lower forms of animal life, the infusorial animalculæ we have already spoken of throw off certain portions, or break themselves up in various directions, sometimes transversely or sometimes longitudinally; or they may give off buds, which detach themselves and develop into their proper forms. There is the common fresh-water polype, for instance, which multiplies itself in this way. Just in the same way as the gardener is able to multiply and reproduce the peculiarities and characters of particular plants by means of cuttings, so can the physiological experimentalist—as was shown by the Abbe Trembley many years ago—so can he do the same thing with many of the lower forms of animal life. M. de Trembley showed that you could take a polype and cut it into two, or four, or many pieces, mutilating it in all directions, and the pieces would still grow up and reproduce completely the original form of the animal. These are all cases of non-sexual multiplication, and there are other instances, and still more extraordinary ones, in which this process takes place naturally, in a more hidden, a more recondite kind of way. You are all of you familiar with that little green insect, the *Aphis* or blight, as it is called. These little animals, during a very considerable part of their existence, multiply themselves by means of a kind of internal budding, the buds being developed into essentially non-sexual animals, which are neither male nor female; they become converted into young *Aphides*, which repeat the process, and their offspring after them, and so on again; you may go on for nine or ten, or even twenty or more successions; and there is no very good reason to say how soon it might terminate, or how long it might not go on if the proper conditions of warmth and nourishment were kept up.

Sexual reproduction is quite a distinct matter. Here, in all these cases, what is required is the detachment of two portions of the parental organisms, which portions we know as the egg or the spermatozoon. In plants it is the ovule and the pollen-grain, as in the flowering plants, or the ovule and the antherozooid, as in the flowerless. Among all forms of animal life, the spermatozoa proceed from the male sex, and the egg is the product of the female. Now, what is remarkable about this mode of reproduction is this, that the egg by itself, or the spermatozoa by themselves, are unable to assume the parental form; but if they be brought into contact with one another, the effect of the mixture of organic substances proceeding from two sources appears to confer an altogether new vigour to the mixed product. This process is brought about, as we all know, by the sexual intercourse of the two sexes, and is called the act of impregnation. The result of this act on the part of the male and female is, that the formation of a new being is set up in the ovule or egg; this ovule or egg soon begins to be divided and subdivided, and to be fashioned into various complex organs, and eventually to develop into the form of one of its parents, as I explained in the first lecture. These are the processes by which the perpetuation of organic beings is secured. Why there should be the two modes—why this re-invigoration should be required on the part of the female element we do not know; but it is most assuredly the fact, and it is presumable, that, however long the process of non-sexual multiplication could be continued—I say there is good reason to believe that it would come to an end if a new

some experimentalists have carefully examined the
lower orders of animals,—among them the Abbé
Spallanzani, who made a number of experiments
upon snails and salamanders,—and have found
that they might mutilate them to an incredible
extent; that you might cut off the jaw or the
greater part of the head, or the leg or the tail, and
repeat the experiment several times, perhaps cutting off the same member again and again; and
yet each of those types would be reproduced
according to the primitive type : Nature making
no mistake, never putting on a fresh kind of leg,
or head, or tail, but always tending to repeat and
to return to the primitive type.

It is the same in sexual reproduction : it is a
matter of perfectly common experience, that the
tendency on the part of the offspring always is,
speaking broadly, to reproduce the form of the
parents.   The proverb has it that the thistle does
not bring forth grapes; so, among ourselves, there
is always a likeness, more or less marked and distinct, between children and their parents.   That is
a matter of familiar and ordinary observation.   We
notice the same thing occurring in the cases of the
domestic animals—dogs, for instance, and their
offspring.   In all these cases of propagation and
perpetuation, there seems to be a tendency in the
offspring to take the characters of the parental
organisms. To that tendency a special name is given
—and as I may very often use it, I will write it

up here on this black-board that you may remember it—it is called *Atavism ;* it expresses this tendency to revert to the ancestral type, and comes from the Latin word *atavus*, ancestor.

Well, this *Atavism* which I shall speak of, is, as I said before, one of the most marked and striking tendencies of organic beings; but, side by side with this hereditary tendency there is an equally distinct and remarkable tendency to variation. The tendency to reproduce the original stock has, as it were, its limits, and side by side with it there is a tendency to vary in certain directions, as if there were two opposing powers working upon the organic being, one tending to take it in a straight line, and the other tending to make it diverge from that straight line, first to one side and then to the other.

So that you see these two tendencies need not precisely contradict one another, as the ultimate result may not always be very remote from what would have been the case if the line had been quite straight.

This tendency to variation is less marked in that mode of propagation which takes place non-sexually ; it is in that mode that the minor characters of animal and vegetable structures are most completely preserved. Still, it will happen sometimes, that the gardener, when he has planted a cutting of some favourite plant, will find, contrary to his expectation, that the slip grows up a little different

from the primitive stock—that it produces flowers
of a different colour or make, or some deviation
in one way or another. This is what is called the
" sporting " of plants.

In animals the phenomena of non-sexual propagation are so obscure, that at present we cannot
be said to know much about them ; but if we turn to
that mode of perpetuation which results from the
sexual process, then we find variation a perfectly
constant occurrence, to a certain extent; and, indeed, I think that a certain amount of variation
from the primitive stock is the necessary result of
the method of sexual propagation itself; for. inasmuch as the thing propagated proceeds from two
organisms of different sexes and different makes
and temperaments, and as the offspring is to be
either of one sex or the other, it is quite clear that
it cannot be an exact diagonal of the two, or it
would be of no sex at all; it cannot be an exact
intermediate form between that of each of its
parents—it must deviate to one side or the other.
You do not find that the male follows the precise
type of the male parent, nor does the female always inherit the precise characteristics of the
mother,—there is always a proportion of the female
character in the male offspring, and of the male
character in the female offspring. That must be quite
plain to all of you who have looked at all attentively
on your own children or those of your neighbours;
you will have noticed how very often it may hap-

pen that the son shall exhibit the maternal type
of character, or the daughter possess the characteristics of the father's family.   There are all sorts of
intermixtures and intermediate conditions between
the two, where complexion, or beauty, or fifty other
different peculiarities belonging to either side of
the house, are reproduced in other members of the
same family.   Indeed, it is sometimes to be remarked in this kind of variation, that the variety
belongs, strictly speaking, to neither of the immediate parents; you will see a child in a family
who is not like either its father or its mother; but
some old person who knew its grandfather or
grandmother, or, it may be, an uncle, or, perhaps,
even a more distant relative will see a great
similarity between the child and one of these.   In
this way it constantly happens that the characteristic of some previous member of the family comes
out and is reproduced and recognised in the most
unexpected manner.

But apart from that matter of general experience,
there are some cases which put that curious mixture in a very clear light.   You are aware that the
offspring of the ass and the horse, or rather of the
he-ass and the mare, is what is called a mule ; and,
on the other hand, the offspring of the stallion
and the she-ass is what is called a hinny.   It is
a very rare thing in this country to see a hinny.
I never saw one myself; but they have been very
carefully studied.   Now, the curious thing is this,

that although you have the same elements in the
experiment in each case, the offspring is entirely
different in character, according as the male influence comes from the ass or the horse. Where the
ass is the male, as in the case of the mule, you
find that the head is like that of the ass, that the
ears are long, the tail is tufted at the end, the feet
are small, and the voice is an unmistakable bray ;
these are all points of similarity to the ass ; but,
on the other hand, the barrel of the body and the
cut of the neck are much more like those of the
mare. Then, if you look at the hinny,—the result
of the union of the stallion and the she-ass, then
you find it is the horse that has the predominance ;
that the head is more like that of the horse, the
ears are shorter, the legs coarser, and the type is
altogether altered ; while the voice, instead of being
a bray, is the ordinary neigh of the horse.   Here,
you see, is a most curious thing : you take exactly
the same elements, ass and horse, but you combine
the sexes in a different manner, and the result is
modified accordingly. You have in this case, however, a result which is not general and universal—
there is usually an important preponderance, but
not always on the same side.

Here, then, is one intelligible, and, perhaps,
necessary cause of variation : the fact, that there
are two sexes sharing in the production of the offspring, and that the share taken by each is different and variable, not only for each combination,

but also for different members of the same family.

Secondly, there is a variation, to a certain extent—though, in all probability, the influence of this cause has been very much exaggerated—but there is no doubt that variation is produced, to a certain extent, by what are commonly known as external conditions,—such as temperature, food, warmth, and moisture. In the long run, every variation depends, in some sense, upon external conditions, seeing that everything has a cause of its own. I use the term "external conditions" now in the sense in which it is ordinarily employed : certain it is, that external conditions have a definite effect. You may take a plant which has single flowers, and by dealing with the soil, and nourishment, and so on, you may by and by convert single flowers into double flowers, and make thorns shoot out into branches. You may thicken or make various modifications in the shape of the fruit. In animals, too, you may produce analogous changes in this way, as in the case of that deep bronze colour which persons rarely lose after having passed any length of time in tropical countries. You may also alter the development of the muscles very much, by dint of training; all the world knows that exercise has a great effect in this way ; we always expect to find the arm of a blacksmith hard and wiry, and possessing a large development of the brachial muscles. No doubt

training, which is one of the forms of external
conditions, converts what are originally only in-
structions, teachings, into habits, or, in other
words, into organisations, to a great extent; but
this second cause of variation cannot be considered
to be by any means a large one.   The third cause
that I have to mention, however, is a very extensive one.   It is one that, for want of a better
name, has been called "spontaneous variation";
which means that when we do not know anything
about the cause of phenomena, we call it spontaneous.   In the orderly chain of causes and
effects in this world, there are very few things of
which it can be said with truth that they are
spontaneous.   Certainly not in these physical
matters—in these there is nothing of the kind—
everything depends on previous conditions.   But
when we cannot trace the cause of phenomena,
we call them spontaneous.

Of these variations, multitudinous as they are,
but little is known with perfect accuracy.   I will
mention to you some two or three cases, because
they are very remarkable in themselves, and also
because I shall want to use them afterwards.
Reaumur, a famous French naturalist, a great
many years ago, in an essay which he wrote upon
the art of hatching chickens—which was indeed a
very curious essay—had occasion to speak of
variations and monstrosities.   One very remarkable case had come under his notice of a variation

in the form of a human member, in the person
of a Maltese, of the name of Gratio Kelleia, who
was born with six fingers upon each hand, and the
like number of toes to each of his feet. That
was a case of spontaneous variation. Nobody
knows why he was born with that number of
fingers and toes, and as we don't know, we call it
a case of "spontaneous" variation. There is
another remarkable case also. I select these,
because they happen to have been observed and
noted very carefully at the time. It frequently
happens that a variation occurs, but the persons
who notice it do not take any care in noting down
the particulars, until at length, when inquiries
come to be made, the exact circumstances are
forgotten; and hence, multitudinous as may be
such "spontaneous" variations, it is exceedingly
difficult to get at the origin of them.

   The second case is one of which you may find
the whole details in the "Philosophical Transactions" for the year 1813, in a paper communicated
by Colonel Humphrey to the President of the
Royal Society—" On a new Variety in the Breed
of Sheep," giving an account of a very remarkable
breed of sheep, which at one time was well known
in the northern states of America, and which
went by the name of the Ancon or the Otter
breed of sheep. In the year 1791, there was a
farmer of the name of Seth Wright in Massachusetts, who had a flock of sheep, consisting of a

ram and, I think, of some twelve or thirteen ewes.
Of this flock of ewes, one at the breeding-time
bore a lamb which was very singularly formed; it
had a very long body, very short legs, and those
legs were bowed. I will tell you by and by how
this singular variation in the breed of sheep came
to be noted, and to have the prominence that it
now has. For the present, I mention only these
two cases; but the extent of variation in the breed
of animals is perfectly obvious to any one who has
studied natural history with ordinary attention, or
to any person who compares animals with others
of the same kind. It is strictly true that there
are never any two specimens which are exactly
alike; however similar, they will always differ in
some certain particular.

Now let us go back to Atavism—to the hereditary tendency I spoke of. What will come of a
variation when you breed from it, when Atavism
comes, if I may say so, to intersect variation?
The two cases of which I have mentioned the
history give a most excellent illustration of what
occurs. Gratio Kelleia, the Maltese, married when
he was twenty-two years of age, and, as I suppose
there were no six-fingered ladies in Malta, he
married an ordinary five-fingered person. The
result of that marriage was four children; the
first, who was christened Salvator, had six fingres
and six toes, like his father; the second was
George, who had five fingers and toes, but one of

them was deformed, showing a tendency to variation; the third was Andrè; he had five fingers and five toes, quite perfect; the fourth was a girl, Marie; she had five fingers and five toes, but her thumbs were deformed, showing a tendency toward the sixth.

These children grew up, and when they came to adult years, they all married, and of course it happened that they all married five-fingered and five-toed persons. Now let us see what were the results. Salvator had four children; they were two boys, a girl, and another boy; the first two boys and the girl were six-fingered and six-toed like their grandfather; the fourth boy had only five fingers and five toes. George had only four children; there were two girls with six fingers and six toes; there was one girl with six fingers and five toes on the right side, and five fingers and five toes on the left side, so that she was half and half. The last, a boy, had five fingers and five toes. The third, Andrè, you will recollect, was perfectly well-formed, and he had many children whose hands and feet were all regularly developed. Marie, the last, who, of course, married a man who had only five fingers, had four children; the first, a boy, was born with six toes, but the other three were normal.

Now observe what very extraordinary phenomena are presented here. You have an accidental variation giving rise to what you may call a monstrosity;

you have that monstrosity or variation diluted
in the first instance by an admixture with
a female of normal construction, and you would
naturally expect that, in the results of such an
union, the monstrosity, if repeated, would be in
equal proportion with the normal type; that is to
say, that the children would be half and half, some
taking the peculiarity of the father, and the others
being of the purely normal type of the mother;
but you see we have a great preponderance of the
abnormal type.   Well, this comes to be mixed once
more with the pure, the normal type, and the abnormal is again produced in large proportion, notwithstanding the second dilution.   Now what
would have happened if these abnormal types had
intermarried with each other; that is to say, suppose the two boys of Salvator had taken it into
their heads to marry their first cousins, the two
first girls of George, their uncle?   You will remember that these are all of the abnormal type of their
grandfather.   The result would probably have been,
that their offspring would have been in every case
a further development of that abnormal type.   You
see it is only in the fourth, in the person of Marie,
that the tendency, when it appears but slightly in
the second generation, is washed out in the third,
while the progeny of Andre, who escaped in the
first instance, escape altogether.

We have in this case a good example of nature's
tendency to the perpetuation of a variation.   Here

it is certainly a variation which carried with it no
use or benefit; and yet you see the tendency to
perpetuation may be so strong, that, notwithstanding a great admixture of pure blood, the variety
continues itself up to the third generation, which
is largely marked with it. In this case, as I have
said, there was no means of the second generation
intermarrying with any but five-fingered persons,
and the question naturally suggests itself, What
would have been the result of such marriage ?
Reaumur narrates this case only as far as the third
generation. Certainly it would have been an exceedingly curious thing if we could have traced this
matter any further; had the cousins intermarried,
a six-fingered variety of the human race might
have been set up.

To show you that this supposition is by no means
an unreasonable one, let me now point out what
took place in the case of Seth Wright's sheep,
where it happened to be a matter of moment to
him to obtain a breed or raise a flock of sheep like
that accidental variety that I have described—and
I will tell you why. In that part of Massachusetts
where Seth Wright was living, the fields were
separated by fences, and the sheep, which were
very active and robust, would roam abroad, and
without much difficulty jump over these fences into other people's farms. As a matter of course,
this exuberant activity on the part of the sheep
constantly gave rise to all sorts of quarrels, bicker-

ings, and contentions among the farmers of the neighbourhood; so it occurred to Seth Wright, who was, like his successors, more or less 'cute, that if he could get a stock of sheep like those with the bandy legs, they would not be able to jump over the fences so readily; and he acted upon that idea. He killed his old ram, and as soon as the young one arrived at maturity, he bred altogether from it. The result was even more striking than in the human experiment which I mentioned just now. Colonel Humphreys testifies that it always happened that the offspring were either pure Ancons or pure ordinary sheep; that in no case was there any mixing of the Ancons with the others. In consequence of this, in the course of a very few years, the farmer was able to get a very considerable flock of this variety, and a large number of them were spread throughout Massachusetts. Most unfortunately, however—I suppose it was because they were so common—nobody took enough notice of them to preserve their skeletons; and although Colonel Humphreys states that he sent a skeleton to the President of the Royal Society at the same time that he forwarded his paper, I am afraid that the variety has entirely disappeared; for a short time after these sheep had become prevalent in that district, the Merino sheep were introduced; and as their wool was much more valuable, and as they were a quiet race of sheep, and showed no tendency to trespass or jump over fences, the Otter

ver might scarcely notice it, and yet every one of
her children has an approximation to the same
peculiarity to some extent. If you look at the
other extreme, too, the gravest diseases, such as
gout, scrofula, and consumption, may be handed
down with just the same certainty and persistence
as we noticed in the perpetuation of the bandy
legs of the Ancon sheep.

However, these facts are best illustrated in
animals, and the extent of the variation, as is well
known, is very remarkable in dogs. For example,
there are some dogs very much smaller than others;
indeed, the variation is so enormous that probably
the smallest dog would be about the size of the
head of the largest; there are very great variations
in the structural forms not only of the skeleton
but also in the shape of the skull, and in the proportions of the face and the disposition of the teeth.

The Pointer, the Retriever, Bulldog, and the
Terrier differ very greatly, and yet there is every
reason to believe that every one of these races
has arisen from the same source,—that all the
most important races have arisen by this selective
breeding from accidental variation.

A still more striking case of what may be done
by selective breeding, and it is a better case, because there is no chance of that partial infusion of
error to which I alluded, has been studied very
carefully by Mr. Darwin,—the case of the domestic
pigeons. I dare say there may be some among you

who may be pigeon *fanciers,* and I wish you to
understand that in approaching the subject, I would
speak with all humility and hesitation, as I regret
to say that I am not a pigeon fancier.   I know it
is a great art and mystery, and a thing upon which
a man must not speak lightly; but I shall endeavour, as far as my understanding goes, to give
you a summary of the published and unpublished
information which I have gained from Mr. Darwin.

Among the enormous variety,—I believe there
are somewhere about a hundred and fifty kinds of
pigeons,—there are four kinds which may be selected as representing the extremest divergences
of one kind from another.   Their names are the
Carrier, the Pouter, the Fantail, and the Tumbler.
In these large diagrams that I have here they are
each represented in their relative sizes to each
other.   This first one is the Carrier; you will
notice this large excrescence on its beak ; it has a
comparatively small head ; there is a bare space
round the eyes ; it has a long neck, a very long
beak, very strong legs, large feet, long wings, and
so on.   The second one is the Pouter, a very large
bird, with very long legs and beak.   It is called
the Pouter because it is in the habit of causing its
gullet to swell up by inflating it with air.   I should
tell you that all pigeons have a tendency to do this
at times, but in the Pouter it is carried to an
enormous extent.   The birds appear to be quite
proud of their power of swelling and puffing them

selves out in this way; and I think it is about as
droll a sight as you can well see to look at a cage
full of these pigeons puffing and blowing themselves out in this ridiculous manner.

This diagram is a representation of the third
kind I mentioned—the Fantail. It is, you see, a
small bird, with exceedingly small legs and a very
small beak. It is most curiously distinguished by
the size and extent of its tail, which, instead of
containing twelve feathers, may have many more,
—say thirty, or even more—I believe there are
some with as many as forty-two. This bird has a
curious habit of spreading out the feathers of its
tail in such a way that they reach forward and
touch its head; and if this can be accomplished, I
believe it is looked upon as a point of great beauty.

But here is the last great variety,—the Tumbler;
and of that great variety, one of the principal
kinds, and one most prized, is the specimen represented here—the short-faced Tumbler. Its beak,
you see, is reduced to a mere nothing. Just compare the beak of this one and that of the first one,
the Carrier—I believe the orthodox comparison of
the head and beak of a thoroughly well-bred Tumbler is to stick an oat into a cherry, and that will
give you the proper relative proportions of the
beak and head. The feet and legs are exceedingly
small, and the bird appears to be quite a dwarf
when placed side by side with this great Carrier.

These are differences enough in regard to their

external appearance; but these differences are by
no means the whole or even the most important of
the differences which obtain between these birds.
There is hardly a single point of their structure
which has not become more or less altered; and to
give you an idea of how extensive these alterations
are, I have here some very good skeletons, for which
I am indebted to my friend, Mr. Tegetmeier, a
great authority in these matters; by means of
which, if you examine them by and by, you will
be able to see the enormous difference in their
bony structures.

I had the privilege, some time ago, of access to
some important MSS. of Mr. Darwin, who, I may
tell you, has taken very great pains and spent
much valuable time and attention on the investigation of these variations, and getting together all
the facts that bear upon them. I obtained from
these MSS. the following summary of the differences between the domestic breeds of pigeons;
that is to say, a notification of the various points
in which their organisation differs. In the first
place, the back of the skull may differ a good deal,
and the development of the bones of the face may
vary a great deal; the back varies a good deal;
the shape of the lower jaw varies; the tongue
varies very greatly, not only in correlation to the
length and size of the beak, but it seems also to
have a kind of independent variation of its own.
Then the amount of naked skin round the eyes,

and at the base of the beak, may vary enormously ;
so may the length of the eyelids, the shape of the
nostrils, and the length of the neck. I have already noticed the habit of blowing out the gullet,
so remarkable in the Pouter, and comparatively so
in the others. There are great differences, too, in
the size of the female and the male, the shape of
the body, the number and width of the processes
of the ribs, the development of the ribs, and the
size, shape, and development of the breastbone.
We may notice, too—and I mention the fact because it has been disputed by what is assumed to
be high authority,—the variation in the number
of the sacral vertebræ. The number of these
varies from eleven to fourteen, and that without
any diminution in the number of the vertebræ of
the back or of the tail. Then the number and
position of the tail-feathers may vary enormously,
and so may the number of the primary and secondary feathers of the wings. Again, the length of
the feet and of the beak,—although they have no
relation to each other, yet appear to go together,—
that is, you have a long beak wherever you have
long feet. There are differences also in the
periods of the acquirement of the perfect plum-
age—the size and shape of the eggs—the nature
of flight, and the powers of flight—so-called
" homing " birds having enormous flying powers ;[1]

[1] The " Carrier," I learn from Mr. Tegetmeier, does not
carry ; a high-bred bird of this breed being but a poor flier.

V

THE CONDITIONS OF EXISTENCE AS AFFECTING THE PERPETUATION OF LIVING BEINGS.

IN the last Lecture I endeavoured to prove to you that, while, as a general rule, organic beings tend to reproduce their kind, there is in them, also, a constantly recurring tendency to vary—to vary to a greater or to a less extent. Such a variety, I pointed out to you, might arise from causes which we do not understand; we therefore called it spontaneous; and it might come into existence as a definite and marked thing, without any gradations between itself and the form which preceded it. I further pointed out, that such a variety having once arisen, might be perpetuated to some extent, and indeed to a very marked extent, without any direct interference, or without any exercise of that process which we called selection. And then I stated further, that by such selection, when exercised artificially—if you took care to breed only from those forms which presented the same peculiarities of any

variety which had arisen in this manner—the
variation might be perpetuated, as far as we can
see, indefinitely.

The next question, and it is an important one
for us, is this: Is there any limit to the amount
of variation from the primitive stock which can
be produced by this process of selective breeding?
In considering this question, it will be useful to
class the characteristics, in respect of which
organic beings vary, under two heads : we may
consider structural characteristics, and we may
consider physiological characteristics.

In the first place, as regards structural characteristics, I endeavoured to show you, by the
skeletons which I had upon the table, and by
reference to a great many well-ascertained facts,
that the different breeds of Pigeons, the Carriers,
Pouters, and Tumblers, might vary in any of their
internal and important structural characters to a
very great degree ; not only might there be changes
in the proportions of the skull, and the characters
of the feet and beaks, and so on ; but that there
might be an absolute difference in the number of
the vertebræ of the back, as in the sacral vertebræ
of the Pouter ; and so great is the extent of the
variation in these and similar characters that I
pointed out to you, by reference to the skeletons
and the diagrams, that these extreme varieties
may absolutely differ more from one another in
their structural characters than do what naturalists

call distinct SPECIES of pigeons; that is to say,
that they differ so much in structure that there is
a greater difference between the Pouter and the
Tumbler than there is between such wild and distinct forms as the Rock Pigeon or the Ring Pigeon,
or the Ring Pigeon and the Stock Dove; and
indeed the differences are of greater value than
this, for the structural differences between these
domesticated pigeons are such as would be admitted by a naturalist, supposing he knew nothing
at all about their origin, to entitle them to constitute even distinct genera.

As I have used this term SPECIES, and shall probably use it a good deal, I had better perhaps devote
a word or two to explaining what I mean by it.

Animals and plants are divided into groups,
which become gradually smaller, beginning with
a KINGDOM, which is divided into SUB-KINGDOMS;
then come the smaller divisions called PROVINCES;
and so on from a PROVINCE to a CLASS, from a
CLASS to an ORDER, from ORDERS to FAMILIES,
and from these to GENERA, until we come at
length to the smallest groups of animals which
can be defined one from the other by constant
characters, which are not sexual; and these are
what naturalists call SPECIES in practice, whatever
they may do in theory.

If, in a state of nature, you find any two groups
of living beings, which are separated one from the
other by some constantly-recurring characteristic,

I don't care how slight and trivial, so long as it is defined and constant, and does not depend on sexual peculiarities, then all naturalists agree in calling them two species; that is what is meant by the use of the word species—that is to say, it is, for the practical naturalist, a mere question of structural differences.[1]

We have seen now—to repeat this point once more, and it is very essential that we should rightly understand it—we have seen that breeds, known to have been derived from a common stock by selection, may be as different in their structure from the original stock as species may be distinct from each other.

But is the like true of the physiological characteristics of animals? Do the physiological differences of varieties amount in degree to those observed between forms which naturalists call distinct species? This is a most important point for us to consider.

As regards the great majority of physiological characteristics, there is no doubt that they are capable of being developed, increased, and modified by selection.

There is no doubt that breeds may be made as different as species in many physiological characters. I have already pointed out to you very

---

[1] I lay stress here on the *practical* signification of " Species." Whether a physiological test between species exist or not, it is hardly ever applicable by the practical naturalist.

briefly the different habits of the breeds of
Pigeons, all of which depend upon their physiological peculiarities—as the peculiar habit of
tumbling, in the Tumbler—the peculiarities of
flight, in the " homing " birds—the strange habit
of spreading out the tail, and walking in a peculiar
fashion, in the Fantail—and, lastly, the habit of
blowing out the gullet, so characteristic of the
Pouter.   These are all due to physiological modifications, and in all these respects these birds differ
as much from each other as any two ordinary
species do.

So with Dogs in their habits and instincts.   It
is a physiological peculiarity which leads the
Greyhound to chase its prey by sight—that enables
the Beagle to track it by the scent—that impels
the Terrier to its rat-hunting propensity—and
that leads the Retriever to its habit of retrieving.
These habits and instincts are all the results of
physiological differences and peculiarities, which
have been developed from a common stock, at
least there is every reason to believe so.   But it
is a most singular circumstance, that while you
may run through almost the whole series of
physiological processes, without finding a check to
your argument, you come at last to a point where
you do find a check, and that is in the reproductive processes.   For there is a most singular circumstance in respect to natural species—at least
about some of them—and it would be sufficient

for the purposes of this argument if it were true
of only one of them, but there is, in fact, a great
number of such cases—and that is, that, similar
as they may appear to be to mere races or breeds,
they present a marked peculiarity in the reproductive process. If you breed from the male and
female of the same race, you of course have offspring of the like kind, and if you make the offspring breed together, you obtain the same result,
and if you breed from these again, you will still
have the same kind of offspring; there is no
check. But if you take members of two distinct
species, however similar they may be to each other,
and make them breed together, you will find a
check, with some modifications and exceptions,
however, which I shall speak of presently. If
you cross two such species with each other, then
—although you may get offspring in the case of
the first cross, yet, if you attempt to breed from
the products of that crossing, which are what are
called HYBRIDS—that is, if you couple a male
and a female hybrid—then the result is that in
ninety-nine cases out of a hundred you will
get no offspring at all ; there will be no result
whatsoever.

The reason of this is quite obvious in some
cases ; the male hybrids, although possessing all
the external appearances and characteristics of
perfect animals, are physiologically imperfect and
deficient in the structural parts of the reproductive

elements necessary to generation. It is said to be invariably the case with the male mule, the cross between the Ass and the Mare; and hence it is, that, although crossing the Horse with the Ass is easy enough, and is constantly done, as far as I am aware, if you take two mules, a male and a female, and endeavour to breed from them, you get no offspring whatever; no generation will take place. This is what is called the sterility of the hybrids between two distinct species.

You see that this is a very extraordinary circumstance; one does not see why it should be. The common teleological explanation is, that it is to prevent the impurity of the blood resulting from the crossing of one species with another, but you see it does not in reality do anything of the kind. There is nothing in this fact that hybrids cannot breed with each other, to establish such a theory; there is nothing to prevent the Horse breeding with the Ass, or the Ass with the Horse. So that this explanation breaks down, as a great many explanations of this kind do, that are only founded on mere assumptions.

Thus you see that there is a great difference between "mongrels," which are crosses between distinct races, and "hybrids," which are crosses between distinct species. The mongrels are, so far as we know, fertile with one another. But between species, in many cases, you cannot succeed in obtaining even the first cross; at any rate it is quite certain that the hybrids are often absolutely infertile one with another.

Here is a feature, then, great or small as it may be, which distinguishes natural species of animals. Can we find any approximation to this in the different races known to be produced by selective breeding from a common stock? Up to the present time the answer to that question is absolutely a negative one. As far as we know at present, there is nothing approximating to this check. In crossing the breeds between the Fantail and the Pouter, the Carrier and the Tumbler, or any other variety or race you may name—so far as we know at present—there is no difficulty in breeding together the mongrels. Take the Carrier and the Fantail, for instance, and let them represent the Horse and the Ass in the case of distinct species; then you have, as the result of their breeding, the Carrier-Fantail mongrel,—we will say the male and female mongrel,—and, as far as we know, these two when crossed would not be less fertile than the original cross, or than Carrier with Carrier. Here, you see, is a physiological contrast between the races produced by selective modification and natural species. I shall inquire into the value of this fact, and of some modifying circumstances by and by; for the present I merely put it broadly before you.

But while considering this question of the limitations of species, a word must be said about what

is called RECURRENCE—the tendency of races
which have been developed by selective breeding
from varieties to return to their primitive type.
This is supposed by many to put an absolute limit
to the extent of selective and all other variations.
People say, " It is all very well to talk about producing these different races, but you know very
well that if you turned all these birds wild, these
Pouters, and Carriers, and so on, they would all return to their primitive stock." This is very commonly assumed to be a fact, and it is an argument
that is commonly brought forward as conclusive ;
but if you will take the trouble to inquire into it
rather closely, I think you will find that it is not
worth very much. The first question of course is,
Do they thus return to the primitive stock ? And
commonly as the thing is assumed and accepted,
it is extremely difficult to get anything like good
evidence of it. It is constantly said, for example,
that if domesticated Horses are turned wild, as
they have been in some parts of Asia Minor and
South America, that they return at once to the
primitive stock from which they were bred. But
the first answer that you make to this assumption
is, to ask who knows what the primitive stock
was ; and the second answer is, that in that case
the wild Horses of Asia Minor ought to be exactly
like the wild Horses of South America. If they
are both like the same thing, they ought mani-
festly to be like each other ! The best authorities,

however, forcibly brought forward by Mr. Darwin, which has been noticed in connection with the breeding of domesticated pigeons; and it is, that however different these breeds of pigeons may be from each other, and we have already noticed the great differences in these breeds, that if, among any of those variations, you chance to have a blue pigeon turn up, it will be sure to have the black bars across the wings, which are characteristic of the original wild stock, the Rock Pigeon.

Now, this is certainly a very remarkable circumstance; but I do not see myself how it tells very strongly either one way or the other. I think, in fact, that this argument in favour of recurrence to the primitive type might prove a great deal too much for those who so constantly bring it forward. For example, Mr. Darwin has very forcibly urged, that nothing is commoner than if you examine a dun horse—and I had an opportunity of verifying this illustration lately while in the islands of the West Highlands, where there are a great many dun horses—to find that horse exhibit a long black stripe down his back, very often stripes on his shoulder, and very often stripes on his legs. I, myself, saw a pony of this description a short time ago, in a baker's cart, near Rothesay, in Bute: it had the long stripe down the back, and stripes on the shoulders and legs, just like those of the Ass, the Quagga, and the Zebra. Now, if we interpret the theory of recurrence as applied to this case, might it not be said that here was a case of a variation exhibiting the characters and conditions of an animal occupying something like an intermediate position between the Horse, the Ass, the Quagga, and the Zebra, and from which these had been developed ? In the same way with regard even to Man. Every anatomist will tell you that there is nothing commoner, in dissecting the human body, than to meet with what are called muscular variations—that is, if you dissect two bodies very carefully, you will probably find that the modes of attachment and insertion of the muscles are not exactly the same in both, there being great peculiarities in the mode in which the muscles are arranged; and it is very singular, that in some dissections of the human body you will come upon arrangements of the muscles very similar indeed to the same parts in the Apes. Is the conclusion in that case to be, that this is like the black bars in the case of the Pigeon, and that it indicates a recurrence to the primitive type from which the animals have been probably developed ? Truly, I think that the opponents of modification and variation had better leave the argument of recurrence alone, or it may prove altogether too strong for them.

To sum up,—the evidence as far as we have gone is against the argument as to any limit to divergences, so far as structure is concerned; and in favour of a physiological limitation. By selective breeding we can produce structural divergences as great as those of species, but we cannot produce equal physiological divergences. For the present I leave the question there.

Now, the next problem that lies before us—and it is an extremely important one—is this: Does this selective breeding occur in nature? Because, if there is no proof of it, all that I have been telling you goes for nothing in accounting for the origin of species. Are natural causes competent to play the part of selection in perpetuating varieties? Here we labour under very great difficulties. In the last lecture I had occasion to point out to you the extreme difficulty of obtaining evidence even of the first origin of those varieties which we know to have occurred in domesticated animals. I told you, that almost always the origin of these varieties is overlooked, so that I could only produce two or three cases, as that of Gratio Kelleia and of the Ancon sheep. People forget, or do not take notice of them until they come to have a prominence; and if that is true of artificial cases, under our own eyes, and in animals in our own care, how much more difficult it must be to have at first hand good evidence of the origin of varieties in nature! Indeed, I do not know that it is possible by direct evidence to prove the origin of a variety in nature, or to prove selective breeding; but I will tell you what we

can prove—and this comes to the same thing—
that varieties exist in nature within the limits of
species, and, what is more, that when a variety has
come into existence in nature, there are natural
causes and conditions, which are amply competent
to play the part of a selective breeder; and although that is not quite the evidence that one
would like to have—though it is not direct testi-
mony—yet it is exceeding good and exceedingly
powerful evidence in its way.

As to the first point, of varieties existing
among natural species, I might appeal to the
universal experience of every naturalist, and of
any person who has ever turned any attention
at all to the characteristics of plants and animals
in a state of nature; but I may as well take
a few definite cases, and I will begin with Man
himself.

I am one of those who believe that, at present,
there is no evidence whatever for saying, that mankind sprang originally from any more than a single
pair; I must say, that I cannot see any good
ground whatever, or even any tenable sort of evidence, for believing that there is more than one
species of Man.   Nevertheless, as you know, just
as there are numbers of varieties in animals, so
there are remarkable varieties of men.   I speak
not merely of those broad and distinct variations
which you see at a glance.   Everybody, of course,
knows the difference between a Negro and a white

man, and can tell a Chinaman from an Englishman. They each have peculiar characteristics of
colour and physiognomy; but you must recollect
that the characters of these races go very far
deeper—they extend to the bony structure, and to
the characters of that most important of all organs
to us—the brain; so that, among men belonging
to different races, or even within the same race,
one man shall have a brain a third, or half, or even
seventy per cent. bigger than another; and if you
take the whole range of human brains, you will
find a variation in some cases of a hundred per
cent. Apart from these variations in the size of
the brain, the characters of the skull vary. Thus
if I draw the figures of a Mongol and of a Negro
head on the blackboard, in the case of the last the
breadth would be about seven-tenths, and in the
other it would be nine-tenths of the total length.
So that you see there is abundant evidence of
variation among men in their natural condition.
And if you turn to other animals there is just the
same thing. The fox, for example, which has a
very large geographical distribution all over
Europe, and parts of Asia, and on the American
Continent, varies greatly. There are mostly large
foxes in the North, and smaller ones in the South.
In Germany alone the foresters reckon some eight
different sorts.

Of the tiger, no one supposes that there is more
than one species; they extend from the hottest

parts of Bengal, into the dry, cold, bitter steppes of Siberia, into a latitude of 50°,—so that they may even prey upon the reindeer. These tigers have exceedingly different characteristics, but still they all keep their general features, so that there is no doubt as to their being tigers. The Siberian tiger has a thick fur, a small mane, and a longitudinal stripe down the back, while the tigers of Java and Sumatra differ in many important respects from the tigers of Northern Asia. So lions vary; so birds vary; and so, if you go further back and lower down in creation, you find that fishes vary. In different streams, in the same country even, you will find the trout to be quite different to each other and easily recognisable by those who fish in the particular streams. There is the same differences in leeches; leech collectors can easily point out to you the differences and the peculiarities which you yourself would probably pass by; so with fresh-water mussels; so, in fact, with every animal you can mention.

In plants there is the same kind of variation. Take such a case even as the common bramble. The botanists are all at war about it; some of them wanting to make out that there are many species of it, and others maintaining that they are but many varieties of one species; and they cannot settle to this day which is a species and which is a variety!

So that there can be no doubt whatsoever that

any plant and any animal may vary in nature;
that varieties may arise in the way I have described
—as spontaneous varieties—and that those varieties may be perpetuated in the same way that I
have shown you spontaneous varieties are perpetuated; I say, therefore, that there can be no doubt
as to the origin and perpetuation of varieties in
nature.

But the question now is :—Does selection take
place in nature? Is there anything like the
operation of man in exercising selective breeding,
taking place in nature? You will observe that,
at present, I say nothing about species; I wish to
confine myself to the consideration of the production of those natural races which everybody
admits to exist. The question is, whether in
nature there are causes competent to produce
races, just in the same way as man is able to produce by selection, such races of animals as we
have already noticed.

When a variety has arisen, the CONDITIONS OF
EXISTENCE are such as to exercise an influence
which is exactly comparable to that of artificial
selection. By Conditions of Existence I mean
two things—there are conditions which are furnished by the physical, the inorganic world, and
there are conditions of existence which are furnished by the organic world. There is, in the first
place, CLIMATE; under that head I include only
temperature and the varied amount of moisture

of particular places. In the next place there is
what is technically called STATION, which means
—given the climate, the particular kind of place
in which an animal or a plant lives or grows; for
example, the station of a fish is in the water, of a
fresh-water fish in fresh water; the station of a
marine fish is in the sea, and a marine animal
may have a station higher or deeper. So again
with land animals : the differences in their stations
are those of different soils and neighbourhoods;
some being best adapted to a calcareous, and
others to an arenaceous soil. The third condition
of existence is FOOD, by which I mean food in
the broadest sense, the supply of the materials
necessary to the existence of an organic being; in
the case of a plant the inorganic matters, such as
carbonic acid, water, ammonia, and the earthy
salts or salines; in the case of the animal the inorganic and organic matters, which we have seen
they require; then these are all, at least the first
two, what we may call the inorganic or physical
conditions of existence. Food takes a mid-place,
and then come the organic conditions; by which
I mean the conditions which depend upon the
state of the rest of the organic creation, upon the
number and kind of living beings, with which an
animal is surrounded. You may class these under
two heads : there are organic beings, which operate
as *opponents*, and there are organic beings which
operate as *helpers* to any given organic creature.

The opponents may be of two kinds : there are
the *indirect opponents*, which are what we may
call *rivals ;* and there are the *direct opponents*,
those which strive to destroy the creature ; and
these we call *enemies*. By rivals I mean, of course,
in the case of plants, those which require for their
support the same kind of soil and station, and,
among animals, those which require the same kind
of station, or food, or climate ; those are the indirect opponents ; the direct opponents are, of
course, those which prey upon an animal or
vegetable. The *helpers* may also be regarded as
direct and indirect : in the case of a carnivorous
animal, for example, a particular herbaceous plant
may, in multiplying, be an indirect helper, by enabling the herbivora on which the carnivore preys
to get more food, and thus to nourish the carnivore
more abundantly ; the direct helper may be best
illustrated by reference to some parasitic creature,
such as the tape-worm. The tape-worm exists in
the human intestines, so that the fewer there are
of men the fewer there will be of tape-worms,
other things being alike. It is a humiliating reflection, perhaps, that we may be classed as direct
helpers to the tape-worm, but the fact is so : we
can all see that if there were no men there would
be no tape-worms.

  It is extremely difficult to estimate, in a proper
way, the importance and the working of the Conditions of Existence. I do not think there were
any of us who had the remotest notion of properly

estimating them until the publication of Mr. Darwin's work, which has placed them before us with remarkable clearness ; and I must endeavour, as far as I can in my own fashion, to give you some notion of how they work. We shall find it easiest to take a simple case, and one as free as possible from every kind of complication.

I will suppose, therefore, that all the habitable part of this globe—the dry land, amounting to about 51,000,000 square miles—I will suppose that the whole of that dry land has the same climate, and that it is composed of the same kind of rock or soil, so that there will be the same station everywhere ; we thus get rid of the peculiar influence of different climates and stations. I will then imagine that there shall be but one organic being in the world, and that shall be a plant. In this we start fair. Its food is to be carbonic acid, water and ammonia, and the saline matters in the soil, which are, by the supposition, everywhere alike. We take one single plant, with no opponents, no helpers, and no rivals ; it is to be a "fair field, and no favour." Now, I will ask you to imagine further that it shall be a plant which shall produce every year fifty seeds, which is a very moderate number for a plant to produce ; and that, by the action of the winds and currents, these seeds shall be equally and gradually distributed over the whole surface of the land. I want you now to trace out what will occur, and you will observe that I am not talking fallaciously any more than a mathematician does when he expounds his problem. If you show that the conditions of your problem are such as may actually occur in Nature and do not transgress any of the known laws of Nature in working out your proposition, then you are as safe in the conclusion you arrive at as is the mathematician in arriving at the solution of his problem. In science, the only way of getting rid of the complications with which a subject of this kind is environed, is to work in this deductive method. What will be the result, then ? I will suppose that every plant requires one square foot of ground to live upon ; and the result will be that, in the course of nine years, the plant will have occupied every single available spot in the whole globe ! I have chalked upon the blackboard the figures by which I arrive at the result :—

Plants.					Plants.
1	× 50 in	1st year	=		50
50	× 50	,, 2nd ,,	=		2,500
2,500	× 50	,, 3rd ,,	=		125,000
125,000	× 50	,, 4th ,,	=		6,250,000
6,250,000	× 50	,, 5th ,,	=		312,500,000
312,500,000	× 50	,, 6th ,,	=		15,625,000,000
15,625,000,000	× 50	,, 7th ,,	=		781,250,000,000
781,250,000,000	× 50	,, 8th ,,	=		39,062,500,000,000
39,062,500,000,000	× 50	,, 9th ,,	=		1,953,125,000,000,000

51,000,000 square miles—the dry surface of the earth × 27,878,400—the number of sq. ft. in 1 sq. mile $\Big\}$ = sq. ft. 1,421,798,400,000,000

being 531,326,600,000,000 square feet less than would be required at the end of the ninth year.

You will see from this that, at the end of the
first year the single plant will have produced fifty
more of its kind ; by the end of the second year
these will have increased to 2,500 ; and so on, in
succeeding years, you get beyond even trillions;
and I am not at all sure that I could tell you what
the proper arithmetical denomination of the total
number really is ; but, at any rate, you will understand the meaning of all those noughts.   Then
you see that, at the bottom, I have taken the
51,000,000 of square miles, constituting the surface of the dry land ; and as the number of square
feet are placed under and subtracted from the
number of seeds that would be produced in the
ninth year, you can see at once that there would
be an immense number more of plants than there
would be square feet of ground for their accommodation.   This is certainly quite enough to
prove my point; that between the eighth and ninth
year after being planted the single plant would have
stocked the whole available surface of the earth.

This is a thing which is hardly conceivable—it
seems hardly imaginable—yet it is so.   It is
indeed simply the law of Malthus exemplified.
Mr. Malthus was a clergyman, who worked out
this subject most minutely and truthfully some
years ago ; he showed quite clearly—and although
he was much abused for his conclusions at the
time, they have never yet been disproved and
never will be—he showed that in consequence of

the increase in the number of organic beings in a
geometrical ratio, while the means of existence
cannot be made to increase in the same ratio, that
there must come a time when the number of organic beings will be in excess of the power of production of nutriment, and that thus some check
must arise to the further increase of those organic
beings.   At the end of the ninth year we have seen
that each plant would not be able to get its full
square foot of ground, and at the end of another
year it would have to share that space with fifty
others the produce of the seeds which it would
give off.

What, then, takes place ?   Every plant grows
up, flourishes, occupies its square foot of ground,
and gives off its fifty seeds ; but notice this, that
out of this number only one can come to anything ;
there is thus, as it were, forty-nine chances to one
against its growing up ; it depends upon the most
fortuitous circumstances whether any one of these
fifty seeds shall grow up and flourish, or whether
it shall die and perish.   This is what Mr. Darwin
has drawn attention to, and called the " STRUGGLE
FOR EXISTENCE " ; and I have taken this simple
case of a plant because some people imagine that
the phrase seems to imply a sort of fight.

I have taken this plant and shown you that this
is the result of the ratio of the increase, the necessary result of the arrival of a time coming for every
species when exactly as many members must be

destroyed as are born; that is the inevitable ultimate result of the rate of production.   Now, what
is the result of all this ?   I have said that there
are forty-nine struggling against every one; and
it amounts to this, that the smallest possible start
given to any one seed may give it an advantage
which will enable it to get ahead of all the others;
anything that will enable any one of these seeds to
germinate six hours before any of the others will,
other things being alike, enable it to choke them
out altogether.   I have shown you that there is
no particular in which plants will not vary from
each other; it is quite possible that one of our
imaginary plants may vary in such a character as
the thickness of the integument of its seeds; it
might happen that one of the plants might produce seeds having a thinner integument, and that
would enable the seeds of that plant to germinate
a little quicker than those of any of the others, and
those seeds would most inevitably extinguish the
forty-nine times as many that were struggling
with them.

I have put it in this way, but you see the practical result of the process is the same as if some
person had nurtured the one and destroyed the
other seeds.   It does not matter how the variation
is produced, so long as it is once allowed to occur.
The variation in the plant once fairly started tends
to become hereditary and reproduce itself; the
seeds would spread themselves in the same way

and take part in the struggle with the forty-nine
hundred, or forty-nine thousand, with which they
might be exposed. Thus, by degrees, this variety
with some slight organic change or modification,
must spread itself over the whole surface of the
habitable globe, and extirpate or replace the other
kinds. That is what is meant by NATURAL
SELECTION; that is the kind of argument by which
it is perfectly demonstrable that the conditions of
existence may play exactly the same part for
natural varieties as man does for domesticated
varieties. No one doubts at all that particular
circumstances may be more favourable for one
plant and less so for another, and the moment you
admit that, you admit the selective power of
nature. Now, although I have been putting a
hypothetical case, you must not suppose that I
have been reasoning hypothetically. There are
plenty of direct experiments which bear out what
we may call the theory of natural selection ; there
is extremely good authority for the statement that
if you take the seed of mixed varieties of wheat
and sow it, collecting the seed next year and sowing it again, at length you will find that out of all
your varieties only two or three have lived, or perhaps even only one. There were one or two
varieties which were best fitted to get on, and they
have killed out the other kinds in just the same
way and with just the same certainty as if you had
taken the trouble to remove them. As I have

sure to be trodden down, crushed, and overpowered by others; and there will be some who just manage to get through only by the help of the slightest accident. I recollect reading an account of the famous retreat of the French troops, under Napoleon, from Moscow. Worn out, tired, and dejected, they at length came to a great river over which there was but one bridge for the passage of the vast army. Disorganised and demoralised as that army was, the struggle must certainly have been a terrible one—every one heeding only himself, and crushing through the ranks and treading down his fellows. The writer of the narrative, who was himself one of those who were fortunate enough to succeed in getting over, and not among the thousands who were left behind or forced into the river, ascribed his escape to the fact that he saw striding onward through the mass a great strong fellow,—one of the French Cuirassiers, who had on a large blue cloak—and he had enough presence of mind to catch and retain a hold of this strong man's cloak. He says, "I caught hold of his cloak, and although he swore at me and cut at and struck me by turns, and at last, when he found he could not shake me off, fell to entreating me to leave go or I should prevent him from escaping, besides not assisting myself, I still kept tight hold of him, and would not quit my grasp until he had at last dragged me through." Here you see was a case of selective saving—if we may

Wyman was there some years ago, and on noticing
no pigs but these black ones, he asked some of the
people how it was that they had no white pigs,
and the reply was that in the woods of Florida
there was a root which they called the Paint
Root, and that if the white pigs were to eat any
of it, it had the effect of making their hoofs crack,
and they died, but if the black pigs ate any of it,
it did not hurt them at all.   Here was a very
simple case of natural selection.   A skilful breeder
could not more carefully develop the black breed
of pigs, and weed out all the white pigs, than the
Paint Root does.

   To show you how remarkably indirect may be
such natural selective agencies as I have referred
to, I will conclude by noticing a case mentioned
by Mr. Darwin, and which is certainly one of the
most curious of its kind.   It is that of the Humble
Bee.   It has been noticed that there are a great
many more humble bees in the neighbourhood of
towns, than out in the open country; and the explanation of the matter is this: the humble bees
build nests, in which they store their honey and
deposit the larvæ and eggs.   The field mice are
amazingly fond of the honey and larvæ; therefore,
wherever there are plenty of field mice, as in the
country, the humble bees are kept down; but in
the neighbourhood of towns, the number of cats
which prowl about the fields eat up the field mice,
and of course the more mice they eat up the less

there are to prey upon the larvæ of the bees—the cats are therefore the INDIRECT HELPERS of the bees.[1] Coming back a step farther we may say that the old maids are also indirect friends of the humble bees, and indirect enemies of the field mice, as they keep the cats which eat up the latter ! This is an illustration somewhat beneath the dignity of the subject, perhaps, but it occurs to me in passing, and with it I will conclude this lecture.

[1] The humble bees, on the other hand, are direct helpers of some plants, such as the heartsease and red clover, which are fertilised by the visits of the bees ; and they are indirect helpers of the numerous insects which are more or less completely supported by the heartsease and red clover.

fessing to discuss a single question, an encyclopædia, I cannot help it.

Now, having had an opportunity of considering in this sort of way the different statements bearing upon all theories whatsoever, I have to lay before you, as fairly as I can, what is Mr. Darwin's view of the matter and what position his theories hold, when judged by the principles which I have previously laid down, as deciding our judgments upon all theories and hypotheses.

I have already stated to you that the inquiry respecting the causes of the phenomena of organic nature resolves itself into two problems—the first being the question of the origination of living or organic beings; and the second being the totally distinct problem of the modification and perpetuation of organic beings when they have already come into existence. The first question Mr. Darwin does not touch; he does not deal with it at all; but he says :—" Given the origin of organic matter—supposing its creation to have already taken place, my object is to show in consequence of what laws and what demonstrable properties of organic matter, and of its environments, such states of organic nature as those with which we are acquainted must have come about." This, you will observe, is a perfectly legitimate proposition; every person has a right to define the limits of the inquiry which he sets before himself; and yet it is a most singular thing that in all the multifarious, and, not unfrequently, ignorant attacks which have been made upon the "Origin of Species," there is nothing which has been more speciously criticised than this particular limitation. If people have nothing else to urge against the book, they say—"Well, after all, you see Mr. Darwin's explanation of the 'Origin of Species' is not good for much, because, in the long run, he admits that he does not know how organic matter began to exist. But if you admit any special creation for the first particle of organic matter you may just as well admit it for all the rest; five hundred or five thousand distinct creations are just as intelligible, and just as little difficult to understand, as one." The answer to these cavils is two-fold. In the first place, all human inquiry must stop somewhere; all our knowledge and all our investigation cannot take us beyond the limits set by the finite and restricted character of our faculties, or destroy the endless unknown, which accompanies, like its shadow, the endless procession of phenomena. So far as I can venture to offer an opinion on such a matter, the purpose of our being in existence, the highest object that human beings can set before themselves, is not the pursuit of any such chimera as the annihilation of the unknown; but it is simply the unwearied endeavour to remove its boundaries a little further from our little sphere of action.

I wonder if any historian would for a moment

admit the objection, that it is preposterous to trouble ourselves about the history of the Roman Empire, because we do not know anything positive about the origin and first building of the city of Rome ! Would it be a fair objection to urge, respecting the sublime discoveries of a Newton, or a Kepler, those great philosophers, whose discoveries have been of the profoundest benefit and service to all men—to say to them—"After all that you have told us as to how the planets revolve, and how they are maintained in their orbits, you cannot tell us what is the cause of the origin of the sun, moon, and stars. So what is the use of what you have done ? " Yet these objections would not be one whit more preposterous than the objections which have been made to the "Origin of Species." Mr. Darwin, then, had a perfect right to limit his inquiry as he pleased, and the only question for us—the inquiry being so limited—is to ascertain whether the method of his inquiry is sound or unsound ; whether he has obeyed the canons which must guide and govern all investigation, or whether he has broken them ; and it was because our inquiry this evening is essentially limited to that question, that I spent a good deal of time in a former lecture (which, perhaps some of you thought might have been better employed), in endeavouring to illustrate the method and nature of scientific inquiry in general. We shall now have to put in practice the principles that I then laid down.

I stated to you in substance, if not in words, that wherever there are complex masses of phenomena to be inquired into, whether they be phenomena of the affairs of daily life, or whether they belong to the more abstruse and difficult problems laid before the philosopher, our course of proceeding in unravelling that complex chain of phenomena with a view to get at its cause, is always the same ; in all cases we must invent an hypothesis; we must place before ourselves some more or less likely supposition respecting that cause ; and then, having assumed an hypothesis, having supposed a cause for the phenomena in question, we must endeavour, on the one hand, to demonstrate our hypothesis, or, on the other, to upset and reject it altogether, by testing it in three ways. We must, in the first place, be prepared to prove that the supposed causes of the phenomena exist in nature ; that they are what the logicians call *vera causæ*— true causes ;—in the next place, we should be prepared to show that the assumed causes of the phenomena are competent to produce such phenomena as those which we wish to explain by them ; and in the last place, we ought to be able to show that no other known causes are competent to produce these phenomena.   If we can succeed in satisfying these three conditions we shall have demonstrated our hypothesis ; or rather I ought to say

we shall have proved it as far as certainty is possible for us; for, after all, there is no one of our
surest convictions which may not be upset, or at
any rate modified by a further accession of knowledge. It was because it satisfied these conditions that we accepted the hypothesis as to the
disappearance of the tea-pot and spoons in the
case I supposed in a previous lecture; we found
that our hypothesis on that subject was tenable
and valid, because the supposed cause existed in
nature, because it was competent to account for
the phenomena, and because no other known cause
was competent to account for them; and it is upon
similar grounds that any hypothesis you choose to
name is accepted in science as tenable and
valid.

What is Mr. Darwin's hypothesis? As I apprehend it—for I have put it into a shape more convenient for common purposes than I could find
verbatim in his book—as I apprehend it, I say,
it is, that all the phenomena of organic nature,
past and present, result from, or are caused by,
the inter-action of those properties of organic
matter, which we have called ATAVISM and VARIA-
BILITY, with the CONDITIONS OF EXISTENCE, or,
in other words,—given the existence of organic
matter, its tendency to transmit its properties, and
its tendency occasionally to vary; and, lastly, given
the conditions of existence by which organic matter is surrounded—that these put together are the

causes of the Present and of the Past conditions of
ORGANIC NATURE.

Such is the hypothesis as I understand it. Now
let us see how it will stand the various tests which
I laid down just now. In the first place, do these
supposed causes of the phenomena exist in nature ?
Is it the fact that, in nature, these properties of
organic matter—atavism and variability—and
those phenomena which we have called the conditions of existence,—is it true that they exist?
Well, of course, if they do not exist, all that I have
told you in the last three or four lectures must be
incorrect, because I have been attempting to prove
that they do exist, and I take it that there is
abundant evidence that they do exist; so far,
therefore, the hypothesis does not break down.

But in the next place comes a much more difficult inquiry:—Are the causes indicated competent to give rise to the phenomena of organic
nature ? I suspect that this is indubitable to a
certain extent. It is demonstrable, I think, as I
have endeavoured to show you, that they are perfectly competent to give rise to all the phenomena
which are exhibited by RACES in nature. Furthermore, I believe that they are quite competent to
account for all that we may call purely structural
phenomena which are exhibited by SPECIES in
nature. On that point also I have already enlarged somewhat. Again, I think that the causes
assumed are competent to account for most of the

physiological characteristics of species, and I not only think that they are competent to account for them, but I think that they account for many things which otherwise remain wholly unaccountable and inexplicable, and I may say incomprehensible. For a full exposition of the grounds on which this conviction is based, I must refer you to Mr. Darwin's work; all that I can do now is to illustrate what I have said by two or three cases taken almost at random.

I drew your attention, on a previous evening, to the facts which are embodied in our systems of Classification, which are the results of the examination and comparison of the different members of the animal kingdom one with another. I mentioned that the whole of the animal kingdom is divisible into five sub-kingdoms; that each of these sub-kingdoms is again divisible into provinces; that each province may be divided into classes, and the classes into the successively smaller groups, orders, families, genera, and species.

Now, in each of these groups the resemblance in structure among the members of the group is closer in proportion as the group is smaller. Thus, a man and a worm are members of the animal kingdom in virtue of certain apparently slight though really fundamental resemblances which they present. But a man and a fish are members of the same sub-kingdom *Vertebrata*, because they are much more like one another than either of them

is to a worm, or a snail, or any member of the other
sub-kingdoms. For similar reasons men and horses
are arranged as members of the same Class, *Mammalia;* men and apes as members of the same
Order, *Primates;* and if there were any animals
more like men than they were like any of the
apes, and yet different from men in important and
constant particulars of their organisation, we should
rank them as members of the same Family, or of
the same Genus, but as of distinct Species.

That it is possible to arrange all the varied
forms of animals into groups, having this sort of
singular subordination one to the other, is a very
remarkable circumstance; but, as Mr. Darwin remarks, this is a result which is quite to be expected, if the principles which he lays down be
correct. Take the case of the races which are
known to be produced by the operation of atavism
and variability, and the conditions of existence
which check and modify these tendencies. Take
the case of the pigeons that I brought before you:
there it was shown that they might be all classed
as belonging to some one of five principal divisions, and that within these divisions other subordinate groups might be formed. The members
of these groups are related to one another in just
the same way as the genera of a family, and the
groups themselves as the families of an order, or
the orders of a class; while all have the same sort
of structural relations with the wild rock-pigeon,

related to the horse indeed. So we may say that animals, in an anatomical sense nearly related to the horse, have those parts which are rudimentary in him fully developed.

Again, the sheep and the cow have no cutting-teeth, but only a hard pad in the upper jaw. That is the common characteristic of ruminants in general. But the calf has in its upper jaw some rudiments of teeth which never are developed, and never play the part of teeth at all. Well, if you go back in time, you find some of the older, now extinct, allies of the ruminants have well-developed teeth in their upper jaws; and at the present day the pig (which is in structure closely connected with ruminants) has well-developed teeth in its upper jaw; so that here is another instance of organs well-developed and very useful, in one animal, represented by rudimentary organs, for which we can discover no purpose whatsoever in another closely allied animal. The whalebone whale, again, has horny " whalebone " plates in its mouth, and no teeth; but the young fœtal whale before it is born has teeth in its jaws; they, how-ever, are never used, and they never come to any-thing. But other members of the group to which the whale belongs have well-developed teeth in both jaws.

Upon any hypothesis of special creation, facts of this kind appear to me to be entirely unaccount-able and inexplicable, but they cease to be so if you accept Mr. Darwin's hypothesis, and see reason for believing that the whalebone whale and the whale with teeth in its mouth both sprang from a whale that had teeth, and that the teeth of the fœtal whale are merely remnants—recollections, if we may so say—of the extinct whale. So in the case of the horse and the rhinoceros : suppose that both have descended by modification from some earlier form which had the normal number of toes, and the persistence of the rudimentary bones which no longer support toes in the horse becomes comprehensible.

In the language that we speak in England, and in the language of the Greeks, there are identical verbal roots, or elements entering into the composition of words. That fact remains unintelligible so long as we suppose English and Greek to be independently created tongues; but when it is shown that both languages are descended from one original, we give an explanation of that resemblance. In the same way the existence of identical structural roots, if I may so term them, entering into the composition of widely different animals, is striking evidence in favour of the descent of those animals from a common original.

To turn to another kind of illustration :—If you regard the whole series of stratified rocks—that enormous thickness of sixty or seventy thousand feet that I have mentioned before, constituting the only record we have of a most prodigious lapse of time, that time being, in all probability, but a fraction of that of which we have no record;—if you observe in these successive strata of rocks successive groups of animals arising and dying out, a constant succession, giving you the same kind of impression, as you travel from one group of strata to another, as you would have in travelling from one country to another;—when you find this constant succession of forms, their traces obliterated except to the man of science —when you look at this wonderful history, and ask what it means, it is only a paltering with words if you are offered the reply—" They were so created."

But if, on the other hand, you look on all forms of organised beings as the results of the gradual modification of a primitive type, the facts receive a meaning, and you see that these older conditions are the necessary predecessors of the present. Viewed in this light the facts of palæontology receive a meaning—upon any other hypothesis I am unable to see, in the slightest degree, what knowledge or signification we are to draw out of them. Again, note as bearing upon the same point, the singular likeness which obtains between the successive Faunæ and Floræ, whose remains are preserved on the rocks : you never find any great and enormous difference between the immediately successive Faunæ and

Floræ, unless you have reason to believe there has also been a great lapse of time or a great change of conditions. The animals, for instance, of the newest tertiary rocks, in any part of the world, are always, and without exception, found to be closely allied with those which now live in that part of the world. For example, in Europe, Asia, and Africa, the large mammals are at present rhinoceroses, hippopotamuses, elephants, lions, tigers, oxen, horses, &c.; and if you examine the newest tertiary deposits, which contain the animals and plants which immediately preceded those which now exist in the same country, you do not find gigantic specimens of ant-eaters and kangaroos, but you find rhinoceroses, elephants, lions, tigers, &c.,—of different species to those now living—but still their close allies. If you turn to South America, where, at the present day, we have great sloths and armadilloes and creatures of that kind, what do you find in the newest tertiaries? You find the great sloth-like creature, the *Megatherium*, and the great armadillo, the *Glyptodon*, and so on. And if you go to Australia you find the same law holds good, namely, that that condition of organic nature which has preceded the one which now exists, presents differences perhaps of species, and of genera, but that the great types of organic structure are the same as those which now flourish.

What meaning has this fact upon any other

hypothesis or supposition than one of successive
modification ? But if the population of the
world, in any age, is the result of the gradual
modification of the forms which peopled it in the
preceding age—if that has been the case, it is intelligible enough ; because we may expect that
the creature that results from the modification of
an elephantine mammal shall be something like
an elephant, and the creature which is produced
by the modification of an armadillo-like mammal
shall be like an armadillo.   Upon that supposition,
I say, the facts are intelligible ; upon any other,
that I am aware of, they are not.

So far, the facts of palæontology are consistent
with almost any form of the doctrine of progressive
modification ; they would not be absolutely inconsistent with the wild speculations of De Maillet,
or with the less objectionable hypothesis of Lamarck.   But Mr. Darwin's views have one peculiar
merit ; and that is, that they are perfectly consistent with an array of facts which are utterly inconsistent with, and fatal to, any other hypothesis
of progressive modification which has yet been
advanced.   It is one remarkable peculiarity of
Mr. Darwin's hypothesis that it involves no necessary progression or incessant modification, and
that it is perfectly consistent with the persistence
for any length of time of a given primitive stock,
contemporaneously with its modifications.   To
return to the case of the domestic breeds of

pigeons, for example; you have the dove-cot
pigeon, which closely resembles the rock pigeon,
from which they all started, existing at the same
time with the others.   And if species are developed
in the same way in nature, a primitive stock and
its modifications may, occasionally, all find the
conditions fitted for their existence; and though
they come into competition, to a certain extent,
with one another, the derivative species may not
necessarily extirpate the primitive one, or *vice
versâ.*

Now palæontology shows us many facts which
are perfectly harmonious with these observed
effects of the process by which Mr. Darwin supposes species to have originated, but which appear
to me to be totally inconsistent with any other
hypothesis which has been proposed.   There are
some groups of animals and plants, in the fossil
world, which have been said to belong to " persist-
ent types," because they have persisted, with
very little change indeed, through a very great
range of time, while everything about them has
changed largely.   There are families of fishes
whose type of construction has persisted all the
way from the carboniferous strata right up to the
cretaceous; and others which have lasted through
almost the whole range of the secondary rocks,
and from the lias to the older tertiaries.   It is
something stupendous this—to consider a genus
lasting without essential modifications through all

this enormous lapse of time while almost everything else was changed and modified.

Thus I have no doubt that Mr. Darwin's hypothesis will be found competent to explain the majority of the phenomena exhibited by species in
nature; but in an earlier lecture I spoke cautiously
with respect· to its power of explaining all the
physiological peculiarities of species.

There is, in fact, one set of these peculiarities
which the theory of selective modification, as it
stands at present, is not wholly competent to
explain, and that is the group of phenomena which
I mentioned to you under the name of Hybridism,
and which I explained to consist in the sterility of
the offspring of certain species when crossed one
with another. It matters not one whit whether
this sterility is universal, or whether it exists only
in a single case. Every hypothesis is bound to
explain, or, at any rate, not be inconsistent with,
the whole of the facts which it professes to account
for; and if there is a single one of these facts
which can be shown to be inconsistent with (I do
not merely mean inexplicable by, but contrary to)
the hypothesis, the hypothesis falls to the ground,
—it is worth nothing. One fact with which it is
positively inconsistent is worth as much, and as
powerful in negativing the hypothesis, as five
hundred. If I am right in thus defining the obli-
gations of an hypothesis, Mr. Darwin, in order to
place his views beyond the reach of all possible

assault, ought to be able to demonstrate the possibility of developing from a particular stock by selective breeding, two forms, which should either be unable to cross one with another, or whose cross-bred offspring should be infertile with one another.

For, you see, if you have not done that you have not strictly fulfilled all the conditions of the problem; you have not shown that you can produce, by the cause assumed, all the phenomena which you have in nature. Here are the phenomena of Hybridism staring you in the face, and you cannot say, "I can, by selective modification, produce these same results." Now, it is admitted on all hands that, at present, so far as experiments have gone, it has not been found possible to produce this complete physiological divergence by selective breeding. I stated this very clearly before, and I now refer to the point, because, if it could be proved, not only that this *has* not been done, but that it *cannot* be done; if it could be demonstrated that it is impossible to breed selectively, from any stock, a form which shall not breed with another, produced from the same stock; and if we were shown that this must be the necessary and inevitable results of all experiments, I hold that Mr. Darwin's hypothesis would be utterly shattered.

But has this been done? or what is really the state of the case? It is simply that, so far as we have gone yet with our breeding, we have not produced from a common stock two breeds which are not more or less fertile with one another.

I do not know that there is a single fact which would justify any one in saying that any degree of sterility has been observed between breeds absolutely known to have been produced by selective breeding from a common stock. On the other hand, I do not know that there is a single fact which can justify any one in asserting that such sterility cannot be produced by proper experimentation. For my own part, I see every reason to believe that it may, and will be so produced. For, as Mr. Darwin has very properly urged, when we consider the phenomena of sterility, we find they are most capricious; we do not know what it is that the sterility depends on. There are some animals which will not breed in captivity; whether it arises from the simple fact of their being shut up and deprived of their liberty, or not, we do not know, but they certainly will not breed. What an astounding thing this is, to find one of the most important of all functions annihilated by mere imprisonment!

So, again, there are cases known of animals which have been thought by naturalists to be undoubted species, which have yielded perfectly fertile hybrids; while there are other species which present what everybody believes to be varieties[1]

_____

[1] And as I conceive with very good reason; but if any objector urges that we cannot prove that they have been produced by

which are more or less infertile with one another. There are other cases which are truly extraordinary; there is one, for example, which has been carefully examined,—of two kinds of sea-weed, of which the male element of the one, which we may call A, fertilises the female element of the other, B; while the male element of B will not fertilise the female element of A; so that, while the former experiment seems to show us that they are *varieties*, the latter leads to the conviction that they are *species*.

When we see how capricious and uncertain this sterility is, how unknown the conditions on which it depends, I say that we have no right to affirm that those conditions will not be better understood by and by, and we have no ground for supposing that we may not be able to experiment so as to obtain that crucial result which I mentioned just now. So that though Mr. Darwin's hypothesis does not completely extricate us from this difficulty at present, we have not the least right to say it will not do so.

There is a wide gulf between the thing you cannot explain and the thing that upsets you altogether. There is hardly any hypothesis in this world which has not some fact in connection with it which has not been explained, but that is a very different affair to a fact that entirely opposes your

artificial or natural selection, the objection must be admitted—ultra-sceptical as it is. But in science, scepticism is a duty.

hypothesis; in this case all you can say is, that your hypothesis is in the same position as a good many others.

Now, as to the third test, that there are no other causes competent to explain the phenomena, I explained to you that one should be able to say of an hypothesis, that no other known causes than those supposed by it are competent to give rise to the phenomena. Here, I think, Mr. Darwin's view is pretty strong. I really believe that the alternative is either Darwinism or nothing, for I do not know of any rational conception or theory of the organic universe which has any scientific position at all beside Mr. Darwin's. I do not know of any proposition that has been put before us with the intention of explaining the phenomena of organic nature, which has in its favour a thousandth part of the evidence which may be adduced in favour of Mr. Darwin's views. Whatever may be the objections to his views, certainly all other theories are absolutely out of court.

Take the Lamarckian hypothesis, for example. Lamarck was a great naturalist, and to a certain extent went the right way to work; he argued from what was undoubtedly a true cause of some of the phenomena of organic nature. He said it is a matter of experience that an animal may be modified more or less in consequence of its desires and consequent actions. Thus, if a man exercise himself as a blacksmith, his arms will become strong and muscular; such organic modification is a result of this particular action and exercise. Lamarck thought that by a very simple supposition based on this truth he could explain the origin of the various animal species: he said, for example, that the short-legged birds which live on fish had been converted into the long-legged waders by desiring to get the fish without wetting their feathers, and so stretching their legs more and more through successive generations. If Lamarck could have shown experimentally that even races of animals could be produced in this way, there might have been some ground for his speculations. But he could show nothing of the kind, and his hypothesis has pretty well dropped into oblivion, as it deserved to do. I said in an earlier lecture that there are hypotheses and hypotheses, and when people tell you that Mr. Darwin's strongly-based hypothesis is nothing but a mere modification of Lamarck's, you will know what to think of their capacity for forming a judgment on this subject.

But you must recollect that when I say I think it is either Mr. Darwin's hypothesis or nothing; that either we must take his view, or look upon the whole of organic nature as an enigma, the meaning of which is wholly hidden from us; you must understand that I mean that I accept it provisionally, in exactly the same way as I accept any other hypothesis. Men of science do not

pledge themselves to creeds; they are bound by
articles of no sort; there is not a single belief that
it is not a bounden duty with them to hold with
a light hand and to part with cheerfully, the
moment it is really proved to be contrary to any
fact, great or small.   And if, in course of time I
see good reasons for such a proceeding, I shall have
no hesitation in coming before you, and pointing
out any change in my opinion without finding the
slightest occasion to blush for so doing.   So I say
that we accept this view as we accept any other,
so long as it will help us, and we feel bound to
retain it only so long as it will serve our great
purpose—the improvement of Man's estate and
the widening of his knowledge.   The moment
this, or any other conception, ceases to be useful
for these purposes, away with it to the four winds;
we care not what becomes of it!

But to say truth, although it has been my business to attend closely to the controversies roused
by the publication of Mr. Darwin's book, I think
that not one of the enormous mass of objections
and obstacles which have been raised is of any
very great value, except that sterility case which
I brought before you just now.   All the rest are
misunderstandings of some sort, arising either
from prejudice, or want of knowledge, or still
more from want of patience and care in reading
the work.

For you must recollect that it is not a book to

be read with as much ease as its pleasant style may lead you to imagine. You spin through it as if it were a novel the first time you read it, and think you know all about it ; the second time you read it you think you know rather less about it ; and the third time, you are amazed to find how little you have really apprehended its vast scope and objects. I can positively say that I never take it up without finding in it some new view, or light, or suggestion that I have not noticed before. That is the best characteristic of a thorough and profound book ; and I believe this feature of the "Origin of Species" explains why so many persons have ventured to pass judgment and criticisms upon it which are by no means worth the paper they are written on.

Before concluding these lectures there is one point to which I must advert—though, as Mr. Darwin has said nothing about man in his book, it concerns myself rather than him ;—for I have strongly maintained on sundry occasions that if Mr. Darwin's views are sound, they apply as much to man as to the lower mammals, seeing that it is perfectly demonstrable that the structural differences which separate man from the apes are not greater than those which separate some apes from others. There cannot be the slightest doub in the world that the argument which applies to the improvement of the horse from an earlier stock, or of ape from ape, applies to the improvement of man from some simpler and lower stock than man. There is not a single faculty—functional or structural, moral, intellectual, or instinctive, there—is no faculty whatever that is not capable of improvement; there is no faculty whatsoever which does not depend upon structure, and as structure tends to vary, it is capable of being improved.

Well, I have taken a good deal of pains at various times to prove this, and I have endeavoured to meet the objections of those who maintain, that the structural differences between man and the lower animals are of so vast a character and enormous extent, that even if Mr. Darwin's views are correct, you cannot imagine this particular modification to take place. It is, in fact, an easy matter to prove that, so far as structure is concerned, man differs to no greater extent from the animals which are immediately below him than these do from other members of the same order. Upon the other hand, there is no one who estimates more highly than I do the dignity of human nature, and the width of the gulf in intellectual and moral matters which lies between man and the whole of the lower creation.

But I find this very argument brought forward vehemently by some. "You say that man has proceeded from a modification of some lower animal, and you take pains to prove that the structural differences which are said to exist in his brain do not exist at all, and you teach that all functions, intellectual, moral, and others, are the expression or the result, in the long run, of structures, and of the molecular forces which they exert." It is quite true that I do so.

"Well, but," I am told at once, somewhat triumphantly, "you say in the same breath that there is a great moral and intellectual chasm between man and the lower animals. How is this possible when you declare that moral and intellectual characteristics depend on structure, and yet tell us that there is no such gulf between the structure of man and that of the lower animals ?"

I think that objection is based upon a misconception of the real relations which exist between structure and function, between mechanism and work. Function is the expression of molecular forces and arrangements no doubt; but, does it follow from this, that variation in function so depends upon variation in structure that the former is always exactly proportioned to the latter ? If there is no such relation, if the variation in function which follows on a variation in structure may be enormously greater than the variation of the structure, then, you see, the objection falls to the ground.

Take a couple of watches—made by the same maker, and as completely alike as possible ; set them upon the table, and the function of each— which is its rate of going—will be performed in the same manner, and you shall be able to distinguish no difference between them; but let me take a pair of pincers, and if my hand is steady enough to do it, let me just lightly crush together the bearings of the balance-wheel, or force to a slightly different angle the teeth of the escapement of one of them, and of course you know the immediate result will be that the watch, so treated, from that moment will cease to go. But what proportion is there between the structural alteration and the functional result? Is it not perfectly obvious that the alteration is of the minutest kind, yet that, slight as it is, it has produced an infinite difference in the performance of the functions of these two instruments?

Well, now, apply that to the present question. What is it that constitutes and makes man what he is? What is it but his power of language—that language giving him the means of recording his experience—making every generation somewhat wiser than its predecessor—more in accordance with the established order of the universe?

What is it but this power of speech, of recording experience, which enables men to be men—looking before and after and, in some dim sense, understanding the working of this wondrous universe—and which distinguishes man from the whole of the brute world? I say that this functional difference is vast, unfathomable, and truly infinite in its consequences; and I say at the same

time, that it may depend upon structural differences which shall be absolutely inappreciable to
us with our present means of investigation. What
is this very speech that we are talking about ?   I
am speaking to you at this moment, but if you
were to alter, in the minutest degree, the proportion of the nervous forces now active in the two
nerves which supply the muscles of my glottis, I
should become suddenly dumb.   The voice is produced only so long as the vocal chords are parallel ;
and these are parallel only so long as certain
muscles contract with exact equality ; and that
again depends on the equality of action of
those two nerves I spoke of.   So that a change of
the minutest kind in the structure of one of these
nerves, or in the structure of the part in which it
originates, or of the supply of blood to that part,
or of one of the muscles to which it is distributed,
might render all of us dumb.   But a race of dumb
men, deprived of all communication with those
who could speak, would be little indeed removed
from the brutes.   And the moral and intellectual
difference between them and ourselves would be
practically infinite, though the naturalist should
not be able to find a single shadow of even specific
structural difference.

But let me dismiss this question now, and, in
conclusion, let me say that you may go away with
it as my mature conviction, that Mr. Darwin's
work is the greatest contribution which has been

made to biological science since the publication of the "Règne Animal" of Cuvier, and since that of the "History of Development," of Von Baer. I believe that if you strip it of its theoretical part it still remains one of the greatest encyclopædias of biological doctrine that any one man ever brought forth ; and I believe that, if you take it as the embodiment of an hypothesis, it is destined to be the guide of biological and psychological speculation for the next three or four generations.

END OF VOL. II

RICHARD CLAY AND SONS, LIMITED,
LONDON AND BUNGAY.